The Development
of Outer Space

The Development
of Outer Space

*Sovereignty and Property Rights
in International Space Law*

Thomas Gangale

PRAEGER

An Imprint of ABC-CLIO, LLC

A B C CLIO

Santa Barbara, California • Denver, Colorado • Oxford, England

Library of Congress Cataloging-in-Publication Data

Gangale, Thomas.
  The development of outer space : sovereignty and property rights in international space law / by Thomas Gangale.
      p. cm.
  Includes index.
  ISBN 978-0-313-37823-2 (hard copy : alk. paper) — ISBN 978-0-313-37824-9 (ebook)
  1. Space law. 2. Moon—International status. 3. Agreement Governing the Activities of States on the Moon and Other Celestial Bodies (1979) I. Title.
  KZD3489.G36 2009
  341.4'7--dc22   2009020515

13 12 11 10 9   1 2 3 4 5

This book is also available on the World Wide Web as an eBook.
Visit www.abc-clio.com for details.

ABC-CLIO, Inc.
130 Cremona Drive, P.O. Box 1911
Santa Barbara, California 93116-1911

This book is printed on acid-free paper ∞
Manufactured in the United States of America

Contents

List of Figures and Tables

FIGURES

TABLES

Preface

Since the culmination of the Apollo program nearly forty years ago, it has been technologically possible to send humans to Mars. Yet we have not done so. Detailed engineering studies in the late 1960s laid out development schedules to support human missions to Mars in the early 1980s. Furthermore, the Apollo program, which could have evolved into extended expeditions to and permanent bases on the Moon, was terminated after only six landing missions. The United States of America retreated from the Moon, and the Union of Soviet Socialist Republics abandoned its effort before achieving success.

I earned my baccalaureate degree in aerospace engineering in the 1970s as the echoes of the Apollo era were fading to silence. By the turn of the 21st century, I began to wonder whether the social forces necessary to take us back to the Moon and further outward to Mars might not yet exist. If they did not exist, to continue dreaming of such projects might be an exercise in futility and frustration. At the very least, in order to put these dreams in their proper context, it would be necessary to understand the social forces that exist today. Also, as the world becomes more integrated economically and politically, it is likely that global economic and political issues will determine when and how we will eventually push off of this little globe into the solar system. Therefore, I turned to the study of international relations as my mission to planet Earth.

Shortly after completing a course in international law, I had a paper accepted for publication in the *Annals of Air and Space Law* refuting the equatorial states' Bogotá Declaration claiming national sovereignty over arcs of the geostationary orbit.[1] During that time the chairman of the Aerospace Architecture Subcommittee (ASASC) of the Design Engineering Technical Committee, American Institute of Aeronautics and Astronautics

(AIAA), posted on the ASASC e-mail list the following message from David Wasser, executive director of the Space Settlement Institute:

> I just want to introduce you to our organization. The Space Settlement Institute believes that the most valuable resource in space is Lunar and Martian real estate. The 1967 Outer Space Treaty prohibits any claims of national sovereignty on the Moon or Mars. But, quite deliberately, the treaty says nothing against private property. Therefore, without claiming sovereignty, the U.S. could recognize land claims made by private companies that establish human settlements there.
>
> We think this is the key to promoting private enterprise in space. Please take a look at our web site.[2]

Having recently acquired some background in international law, and, in particular, having spent some time researching the Outer Space Treaty, my immediate impression of Wasser's assertion was that it was almost certainly wrong. I took him up on his invitation, and as I looked into the Space Settlement Institute's online literature, it became clear to me that the institute's fundamental purpose was contrary to established principles of international law in general as well as to the Outer Space Treaty specifically. On January 28, 2005, I responded:

> As a matter of natural law theory, rights exist independently of their legal recognition. As Jefferson wrote, we "are endowed . . . with certain unalienable rights." Governments are instituted to secure these rights. The "pursuit of Happiness" has been interpreted as an 18th-century poetic phrase meaning "property rights." As a practical matter, however, rights to a specific property exist only if they are granted or recognized by a government and subject to the protection of law. Such grant, recognition, or protection is an act of sovereignty. Title cannot come into existence out of thin air (or the vacuum of space). Legal title must arise from a sovereign power possessing legal authority over the territory in question.
>
> Referring to the Space Settlement Institute's "Land Claim Recognition (LCR) Analysis" at http://www.space-settlement-institute.org/Articles/LCRbrieftext.htm, "Congress should pass 'land claim recognition' legislation legalizing private claims of land in space. A land claim recognition bill would not violate the ban on sovereign ownership if the 'use and occupation' standard from civil law (rather than 'gift of the sovereign' from common law) were used as the legal basis for the private claim."
>
> This is flat out wrong.
>
> Article II of the 1967 Outer Space Treaty states: "Outer space, including the Moon and other celestial bodies, is not subject to national appropriation by claim of sovereignty, by means of use or occupation, or by any other means." Article VI of the Outer Space Treaty provides: "States Parties to the Treaty shall bear international responsibility for national activities in outer

space, including the moon and other celestial bodies, whether such activities are carried on by governmental agencies or by non-governmental entities, and for assuring that national activities are carried out in conformity with the provisions set forth in the present Treaty. The activities of non-governmental entities in outer space, including the moon and other celestial bodies, shall require authorization and continuing supervision by the appropriate State Party to the Treaty."

"Non-governmental entities" includes private parties; thus States Parties cannot appropriate the Moon and other celestial bodies, or parts thereof, through private parties.

For Congress to pass "land claim recognition" legislation legalizing private claims of land in space would be an exercise of sovereignty, and therefore a violation of international law under the provisions of the 1967 Outer Space Treaty. Although I personally favor the development of an international legal regime for granting and protecting extraterrestrial property rights, I also deplore unilateral action on the part of the United States as an international outlaw. In my view, we have seen far too much of that in the past four years.[3]

I never heard from Dave Wasser again. However, more than a month later, on March 5, 2005, I heard from his father:

My name is Alan Wasser, and I am the Chairman of the Space Settlement Institute as well as a member of the AIAA Space Colonization Technical Committee (SCTC). I would like the opportunity to address your concerns.

As you know, the Institute is convinced that land claims recognition legislation would create the ultimate economic incentive for private industry (and not just private industry in the U.S. by the way) to establish a permanent space settlement on the Moon—and that it would be legal to do so. No other plan even comes close as an incentive, and I will show why.[4]

At this point, the elder Wasser, a former chairman of the Executive Committee of the National Space Society, launched into specious real estate pitch, using cherry-picked land value data, in an effort to convince me that land in Nevada is quite valuable and that land on the Moon is even more so. However, since the present work examines outer space law, not exotic real estate ventures, this section of Wasser's message can be omitted. He then turned to the legal issues. Regarding my statement that "land claim recognition" legislation would be an exercise of sovereignty, and therefore a violation of the Outer Space Treaty:

That is your opinion—and you are certainly not alone in that opinion.

But there are others of a different opinion, and, in this kind of international law, there is no judge, no court, qualified to give us a binding ruling.

Therefore, in the end, all that really matters is the opinion of the U.S. Congress. If it decides it would NOT be an exercise of sovereignty, then it would not be an exercise of sovereignty.[5]

Who are these "others of a different opinion" to whom Wasser refers? In this work I will explore at length both the legality and the wisdom, in terms of policy implications, of Wasser's position and those of the "others." Regarding real property rights, a number of space lawyers around the world, space advocacy organizations, and at least one astronaut and one member of Congress, have weighed in. A number of public statements by prominent figures openly challenge the system of international outer space law as being an obstruction to the commercial development of extraterrestrial resources, particularly the Moon and Mars, and go as far as advocating abrogation of the 1967 Outer Space Treaty by the United States. And, one writer on space law seemed to froth at the mouth at my very mention of the 1979 Moon Agreement: "It's dead, dead, dead!" There will be other issues of sovereignty and property in outer space to explore as well, but even these to some extent will cast a light of scrutiny on the character of the Space Settlement Institute's aims and methods. But for now, consider Wasser's following declaration:

> Of course, there will be dissenters who will say they deplore such unilateral action on the part of the United States. Perhaps they'll also say, in their view, the U.S. is an international outlaw.
> But the reward will be of immense benefit to all mankind by expanding the habitat of humanity and opening the space frontier for all.[6]

Similarly, C. S. Lewis's character in *Out of the Silent Planet*,[7] Professor Edward Rolles Weston, asserts that whatever serves the great cause of expanding the human race across the cosmos is moral. In other words, the ends will justify the means. How often that excuse has been used throughout history, often with devastating consequences, whether intended or unintended. We are invited to dismantle a legal structure that has existed for 40 years . . . and erect what in its place? President John F. Kennedy made the following remarks at the dawn of human spaceflight:

> We set sail on this new sea because there is new knowledge to be gained and new rights to be won, and they must be won and used for the progress of all people. For space science, like nuclear science and all technology, has no conscience of its own. Whether it will become a force for good or ill depends on man, and only if the United States occupies a position of pre-eminence can we help decide whether this new ocean will be a sea of peace or a new terrifying theater of war. I do not say the we should or will go unprotected

against the hostile misuse of space any more than we go unprotected against the hostile use of land or sea, but I do say that space can be explored and mastered without feeding the fires of war, without repeating the mistakes that man has made in extending his writ around this globe of ours.[8]

It must be considered that where the rule of law does not exist, there can be no rights, either old or new, whether for the progress of all people or for the privilege of the few. If we are to discard behind us here on Earth certain principles, shall we do so out of mere expediency, or because we find new, higher, far-reaching ones to replace them? We must choose the higher road to the High Frontier, or be brought low by our baser motivations.

Fictional literature is at its best when it explores the human condition, when it asks questions as to what it means to be human: what is love, what is life, what is justice, what is truth? Accordingly, science fiction also attains its greatest height when it explores the same questions, but uses exotic settings and situations to ask them in new ways and to shine a different light on them. One of the best episodes of the dramatic television series *Star Trek: The Next Generation*, titled "The Measure of a Man," explores the question of whether an android has the same rights of legal personality as any organic sentient being. In summing up his case for the android Data, starship Captain Jean-Luc Picard declares to the judge:

> The decision you reach here today will determine how we will regard this creation of our genius. It will reveal the kind of a people we are, what he is destined to be. It will reach far beyond this courtroom and this one android. It could significantly redefine the boundaries of personal liberty and freedom, expanding them for some, savagely curtailing them for others. Are you prepared to condemn him and all who come after him to servitude and slavery? Your honor, Starfleet was founded to seek out new life. Well, there it sits . . . waiting. You wanted a chance to make law, well here it is. Make it a good one.[9]

And so the stage is set, not for a futuristic, cosmic morality play, but for a serious inquiry into what moral and legal principles we will carry into the future and into the cosmos. What challenges actually await us in outer space are far beyond what we can imagine, even now, half a century into the Space Age, for in that time, except for nine sprints to the Moon and back, crews have streaked across the skies just a few hundred kilometers above the heads of the Earthbound. This is to interplanetary spacefaring what coastal canoe fishing is to intercontinental shipping. The challenges that await us are not technological alone. The science fiction novelist Robert A. Heinlein wrote that the Moon is a harsh mistress.[10] All of the new worlds—not just the Moon—will be harsh mistresses. We will live close to the edge of extinction out there, but learning to survive on those other worlds will bring our species closer to immortality. We will learn to depend on each other for our very lives as never before—Africans, Americans, Asians, Australians, Europeans—all of us. The vastness that Kennedy called "the New Frontier" will be punctuated by claustrophobic habitat modules, not sprawling with the wide-open spaces of the American Old West. We will live in enclosed places, in each other's faces. All the pretentious barriers that we erect here on Earth will melt away in space. We will come to know each other—and ourselves—as we have never done before. We will push the outside of the envelope of what it means to be human. Living together so closely, so intimately, so inescapably, will tear down social and psychological walls that we need not and dare not consider here on our comfortable, capacious, suburbanized, subdivided Earth. There will be new challenges to human dignity, privacy, individuality, intimacy, and polity, and these will be rich fields of inquiry to the social sciences.

And we will need to make new law for all the things that we discover to be human, including those new things that we may fashion from our ingenuity and those new things that we will discover within ourselves. Let us make it a good one.

NOTES

1. Gangale, Thomas. 2006. "Who Owns the Geostationary Orbit?" *Annals of Air and Space Law.* XXXI:425–446.

2. Wasser, David. 2005. E-mail, January 26.

3. Gangale, Thomas. 2005. E-mail, January 28.

4. Wasser, Alan. 2005. E-mail, March 5.

5. Ibid.

6. Ibid.

7. Lewis, C. S. 1938. *Out of the Silent Planet.* London: John Lane.

8. Kennedy, John F. 1962. "Address at Rice University on the Nation's Space Effort." September 12. Available from http://www.jfklibrary.org/j091262.htm; accessed December 30, 2005.

9. Snodgrass, Melinda M. 1989. "The Measure of a Man," Act 5, Scene 22. *Star Trek: The Next Generation.* Produced by Gene Roddenberry, directed by Robert Scheerer. 46 min. Paramount Pictures. DVD and VHS.

10. Heinlein, Robert A. 1966. The Moon Is a Harsh Mistress. New York: G. P. Putnam.

1

The Forsaken Promise of Space

The golden age of space exploration was a product of the Cold War. Space was simply another front in that war. The United States and the Soviet Union had strategic and public diplomacy interests in space. In this heady atmosphere, Arthur C. Clarke published *The Promise of Space* in 1968, foretelling the wonders yet to come.[1] These ideas, popularized by Clarke and later by Gerard K. O'Neill in 1977,[2] were all the rage in the 1970s in the nascent community of space enthusiasts and provided the momentum for the creation of the National Space Association in 1974 (renamed the National Space Institute the following year) and the L-5 Society in 1975 (the two organizations merged to become the National Space Society in 1987). As director of Niagara University's Center for the Study of Human Communities in Space (later renamed the Space Settlement Studies Project), Stewart B. Whitney* wrote:

> Space industrialization promises to alleviate many of earth's problems and to help solve the calamity of despoiled habitation on this planet. It is a new growth area to stabilize the business environment, to stimulate investment, and to help fashion a new political economy. The manufacture and utilization of resources located off earth for the benefit of new and improved products, services and sources of energy for earth will profit those who possess the imagination and resources to exploit this new frontier. Economic self-interest and public good will generate investment in space industrialization to produce huge dividends including financial returns, scientific knowledge, development of new industry and products, an alternative solution to

---

* Whitney was slated to be the third "Teacher in Space" at the time of the Space Shuttle OV-099 *Challenger* STS-51L accident in January 1986, which carried the first "Teacher in Space," Christa McAuliffe.

world problems of resource depletion, environmental pollution, and additional habitat space.[3]

Four decades have passed since Clarke's *The Promise of Space*, with these visions still unrealized. Still, many continue to see this as the inevitable, although delayed, future. According to Wayne N. White, an author on space law,

> Professionals foresee an integrated system of solar power generation, lunar and asteroidal mining, orbital industrialization, and habitation in outer space.[4]

Although the activities listed by White are still being discussed, not a single one of them has come into being. Nevertheless, he insists:

> A development regime which provides some form of property rights will become increasingly necessary as space develops.[5]

The statement is easily accepted, but the crucial question is, when will an actual regime of even limited scope become necessary? At this time, humankind is barely capable of sustaining an astronaut and a cosmonaut aboard the International Space Station right above our heads, which is a far cry from O'Neill's far-flung space colonies housing hundreds of thousands. By a number of important technological measures, humankind is further from exploiting the Moon and other celestial bodies than it was in 1970. At that time, the United States had the Saturn V launch vehicle, which was capable of sending 47,000 kg to the Moon. It also possessed the Nuclear Engine for Rocket Vehicle Applications (NERVA), which, as the basis for a new upper stage for the Saturn V, could have increased the Saturn V translunar capability to 160,000 kg.[6] Also at that time, the Soviet Union was developing the N1 launch vehicle, which was less capable than the Saturn V but exceeded the payload capability of any launch vehicle that has been commercially developed (see Table 1.1 and Figure 1.1). None of these systems exist today, nor are comparable systems likely to be developed by commercial enterprises in the foreseeable future.* Absent such systems, talk of "an integrated system of solar power generation, lunar and asteroidal mining, orbital industrialization, and habitation in outer space" remains wildly premature.

---

* The most recent comparable system was the Soviet Energiya launch vehicle, which was capable of lifting 22,000 kg to a geosynchronous transfer orbit. The Energiya had only two flights, in May 1987 and November 1988 (http://www.astronautix.com/lvs/energia.htm). The United States plans to launch its Ares V launcher in 2018.

Table 1.1: Payload Capabilities of Heavy-Lift Launch Vehicles

Launch Vehicle	Nation	Last Launch Date	Low Earth Orbit Payload (kg)	Escape Trajectory Payload (kg)
Delta 7000H	USA	August 3, 2004		1,000
Ariane 42L	France	September 25, 2001	7,600	3,590
Atlas V	USA	December 17, 2004	12,500	5,000
H-2A	Japan	November 29, 2003	11,730	5,000
N1	USSR	November 23, 1972	75,000	20,000
Energiya	USSR	November 15, 1988	80,000	22,000
Saturn V	USA	May 14, 1973	116,000	47,000
Saturn V/NERVA	USA	(1981)		160,000

Source: www.astronautix.com

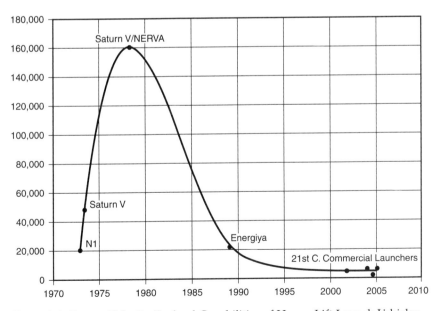

Figure 1.1: Escape Velocity Payload Capabilities of Heavy-Lift Launch Vehicles

Given these limitations, why all of the clamor over property rights in outer space? When White contends that "the right to maintain a facility in a given location relative to another space object may create . . . conflicts [which] may arise sooner than we expect," one ought to bear in mind that we have been expecting for thirty years, and that has been one very long gestation.

There have been a couple of important sociopolitical reactions to the forsaken promise of space. In the developed world, a libertarian movement argues for government—all governments—to get out of the way so entrepreneurs can usher in the "real" space age. In the developing world, some think the current international legal regime is failing to deliver on the long expressed principle that the launching states share the adventure and benefits with the developing states, bringing them more rapidly into the international partnership of space development.

In the United States—the nation that walked on the Moon, operated the first partially reusable space transportation system, and explored all of the planets in the Solar System—some space enthusiasts have concluded that government dropped the ball when it came to space development. Moreover, they have concluded that government has blocked others from picking up the ball and running with it. As the Apollo era came to a close, many anticipated a *fully* reusable launch system, a permanently occupied space station, a Moon base, and human expeditions to Mars—all by the early 1980s. The necessary technology was either in hand or within reach, so why did none of this happen? Pointing the finger at a convenient scapegoat, frustrated space enthusiasts have adopted the libertarian mantra that "government is the problem," and refer not only to the U.S. government, but also to national governments in general and the international treaty regime they have created to govern outer space.

> We are currently saddled with a legal framework that was conceived and created at the height of the Cold War. The Outer Space Treaty of 1967 and its progeny established a basis for the allocation of property rights in outer space, subsumed under the concept of the "common heritage of mankind" . . . that is antithetical to the economic development of space resources. Interpretations of the Outer Space Treaty's rules regarding the appropriation of both land and resources are varied. One camp argues that the treaty only prohibits national appropriation, not private appropriation. While others argue that private appropriation must be endorsed by a state to survive; this state endorsement is then interpreted as a form of state appropriation, and thus, private appropriation is clearly prohibited by the treaty.
>
> As a result of this confusion, progress has been considerably slower than it might have otherwise been. While there are many private and publicly funded companies with plans ranging from landing microspacecrafts [sic] on asteroids to developing a lunar rover and collecting samples of lunar soil for eventual sale on Earth, none are moving ahead quickly in light of the uncertain property laws.[7]

This critique ignores the question of whether there was ever a viable economic rationale for the envisioned space wonders of the 1980s. In any case, the libertarian space agenda is promoting ownership rights to extra-terrestrial resources and real property in accordance with current international law where possible, by modifying international law if feasible, or by destroying the international treaty regime if necessary.

Alan Wasser, former chairman of the Executive Committee of the National Space Society, has proposed federal legislation mandating that the United States recognize extraterrestrial claims to real property, based on a unilateral reinterpretation of the 1967 Outer Space Treaty. The Lunar Settlement Initiative espoused by the Lunar Republic Society is in the same vein but of lesser import. Nevertheless, the issue of private property rights in outer space has been one of high visibility in the American citizen pro-space movement for more than a quarter-century, and the issue has profound policy ramifications in both the short term (because the proposal is a current subject of debate) and the long term (because, if enacted, it would a have dramatic effect on international space law and would shape the future commercial development of outer space).

There are even more specious schemes and scams on the New Frontier in the name of private property rights. What is the case for or against the current international legal regime? Has it indeed impeded space development? Are there certain claims of national sovereignty and private property that can be asserted under the present international law of outer space? If not, should changes be made to permit such claims? In what directions should the international law of outer space develop? This book reaches two main conclusions: First, current international law is not the barrier to the economic development of outer space; technology is the barrier and will remain a significant barrier to private enterprise for a long time to come. Second, the libertarian agenda to tear down the international legal regime of outer space is based on the erroneous premise that international law presents a barrier to private enterprise, and the success of this agenda would create legal uncertainty, thereby discouraging investment and setting the stage for armed conflict in outer space.

To begin, we look at how space law developed from terrestrial law: theories on property from the Enlightenment, space law's roots in maritime and aviation law, and an overview of the United Nations' declarations and international agreements. Next we explore the place property rights currently occupies in space law (particularly under the Outer Space Treaty), critics of the treaty, and innovative ideas to strengthen the legal protection of property. Then we consider the Moon Agreement: the story of its defeat, an examination of the arguments against it, and the case for reviving it as the United States prepares to resume the human exploration of the Moon. Having explored the Outer Space Treaty and the Moon

Agreement, we look at how the Space Settlement Initiative turns international law on its head, as well as how it relies on a distorted rewriting of history, misunderstanding of legal principles, and a new adventure in voodoo economics.

After a defense of the existing international legal regime of outer space, we explore its future implications. The latter third of the book begins by considering the rise of China as a space power and the relative decline of the United States. Is China an expansionist power? Would it be wise for the United States to abandon the Outer Space Treaty and thereby tear down an international legal regime that precludes a major motivation for armed conflict—territorial claims—or is it better off embedding itself and China in the stability of international institutions? Next, we speculate on the evolution of an interplanetary political economy: the tension between technocracy and technoeconomy, the need for a management of the commons (as most of space will remain under the existing legal paradigm, even as we settle it), the relative advantages of planets in the Solar System economy, and the possible transformation of the Westphalian nation-state system and capitalism. Finally, we wrestle with a question that still resonates with many who so many years ago watched with fascination as astronauts on the Moon were broadcast on our television screens for hours on end: what political, economic, and social factors might trigger the next great age of human exploration? This and the other questions addressed in this book are important not because they help us reconcile with a vision of the future that did not come to pass, but because although far more awaits us in outer space than we have attained, there are not only technical and physical risks in how we reach for the stars, but there are also dangers to American political ideals and economic principles right here on Earth.

NOTES

1. Clarke, Arthur C. 1968. *The Promise of Space.* New York: Harper & Row.

2. O'Neill, Gerard K. 1977. *The High Frontier: Human Colonies in Space.* New York: Morrow.

3. Whitney, Stewart B. 1984. "Space Political Economy: Integrating Technology and Social Science for the 1990s." Third Annual Space Development Conference, San Francisco, April.

4. White, Wayne N. 1998. "Real Property Rights in Outer Space." *Proceedings, 40th Colloquium on the Law of Outer Space,* 370. American Institute of Aeronautics and Astronautics. Available from http://www.spacefuture.com/archive/real_property_rights_in_outer_space.shtml; accessed March 19, 2005.

5. Ibid.

6. Wade, Mark. [n.d.]. "Saturn V." *Encyclopedia Astronautica.* Available from http://astronautix.com/lvs/saturnv.htm; accessed October 23, 2008.

7. Fountain, Lynn M. 2003. "Creating Momentum in Space: Ending the Paralysis Produced by the 'Common Heritage of Mankind' Doctrine." *Connecticut Law Review,* 35:1753. Available from http://0-web.lexis-nexis.com.opac.sfsu.edu/universe/document?_m=11fc26b13c17a600a2aaf5b31ce697fb&_docnum=28&wchp=dGLbVtb-zSkVb&_md5=2b3f6fb9a388d7b1b4b0f72e7eebeda1; accessed January 25, 2006.

2

The Launching of
Space Law

SOVEREIGNTY AND PROPERTY ON EARTH

John Locke wrote of material property rights as arising from the mixing of a person's labor (which he owns intrinsically) with the natural substance of the Earth:

> Though the Earth, and all inferior creatures, be common to all men, yet every man has a property in his own person: this no body has any right to but himself. The labour of his body, and the work of his hands, we may say, are properly his. Whatsoever then he removes out of the state that nature hath provided, and left it in, he hath mixed his labour with, and joined to it something that is his own, and thereby makes it his property. It being by him removed from the common state nature hath placed it in, it hath by this labour something annexed to it, that excludes the common right of other men: for this labour being the unquestionable property of the labourer, no man but he can have a right to what that is once joined to, at least where there is enough, and as good, left in common for others.[1]

From the beginning of international law on outer space, the issues of property have been intertwined with those of sovereignty. The reason for this is fairly straightforward. Although natural law theorists such as Locke may argue that rights exist by virtue of human nature and merely await discovery by law and society, governments necessarily deal in rights that they have identified and have agreed to recognize. Thomas Hobbes argued that the rights of the subjects of a commonwealth include the right to engage in lawful commerce and the right to legal protection of their property.[2]

A property right must be recognized by a sovereign government for that right to have legal force. As a practical matter, a property right cannot exist in the absence of a controlling legal regime. Outside such a regime, where a state of anarchy prevails, any claim to property must be defended by the force of arms; it is not a right, but a physical fact of occupation.

The concept of sovereignty itself is similarly constructed by the community of nations. A state is sovereign if it is generally recognized as such by other sovereign states. It need not possess the capacity to physically defend its claim of sovereignty (Iceland, for instance, has no military establishment), but may be guaranteed sovereignty by security alliances (such as the North Atlantic Treaty Organization) or by international institutions that seek to regulate state interaction (such as the United Nations). Absent international recognition, a government must rely to a large degree on self-help, defending its claim of sovereignty by the threat of force. In such a case, the right of sovereignty is unrecognized, yet sovereignty is a physical fact. Usually this is a transitional phase, leading either to ultimate victory or final defeat. For example, the United States of America exercised sovereignty over much, although not all, of its claimed territory between declaring independence in 1776 and receiving general international recognition in 1783. The Confederate States of America (1861–1865), on the other hand, exercised sovereignty but was never recognized. Rhodesia (1965–1979) and Transnistria (1991–present) are among the longest-running examples of unrecognized, *de facto* sovereignty. The Republic of China (1912–present), which has continued to exercise sovereignty over Taiwan since being forced to withdraw from mainland China in 1949, is a case of a long-recognized government being gradually de-recognized by the international community. The ROC was a founding member of the United Nations, yet its seat was given to the People's Republic of China in 1971. Most states switched their recognition from the ROC to the PRC in the 1970s, and at this time, the ROC is officially recognized by only 25 Third World states, although it maintains unofficial relations with most major powers.

SOVEREIGNTY AND PROPERTY IN OUTER SPACE

A number of early writers recognized the absurdity of national sovereignty reaching to infinity, because in such a system a spacecraft would constantly pass through the legal jurisdiction of one subjacent state after another. However, this question was not addressed in the 1944 Convention on International Civil Aviation. A. G. Haley asserted that, in theory, national sovereignty should extend as far as necessary to assure or protect the state; however, this did not lead to a practical definition.[3] In 1951, John. C. Cooper proposed that sovereignty be extended as far as Earth's gravitational influence.[4] One trouble with this idea is that the parameters are difficult to define precisely. Technically speaking, Earth's gravitational influence extends to infinity, diminishing as the square of the distance increases. Certainly, the Moon, orbiting around the Earth, is within the Earth's gravitational influence, but what of an object in a Moon-intersecting trajectory, in orbit around, or on the surface of the Moon? Also, more distant objects than the Moon are materially influenced by the Earth's gravity, such as 3753 Cruithne, a five-kilometer asteroid that traces a very complex path in relation to the Earth and the Sun and is locked into a solar orbital period exactly synchronized with Earth's. Since its discovery in 1986, 3753 Cruithne has been called our second moon, yet at times it is 380 million kilometers from Earth.*

Noted scholars Oscar Schachter, A. Meyer, and C. Wilfred Jenks believed that sovereignty ends where airspace ends.[5] Meyer reasoned that states have sovereignty over airspace only because the atmosphere is part of the Earth's environment, and beyond the atmosphere—outer space— is a separate legal environment.

Beginning in the early 1950s, the United States desired outer space to be internationally recognized as a commons. This was long before it became apparent that the Soviet Union had an initial advantage in space launch capability and might by the first to develop the capability of reaching other planets. At this time, the United States viewed outer space in terms of its own security. The Soviet Union had emerged from the World War II as a hostile superpower, and strategic intelligence was a vital component of the developing Cold War. As an open society with a free press and a mobile population, the United States was a soft target for the Soviet intelligence services; on the other hand, the secretiveness of Soviet society, with its government-controlled press and its travel restrictions on citizens and foreign visitors alike, made the collection of intelligence by technical means (e.g., surveillance from above Soviet territory) a priority for the United States.

In the late 1940s, it was already clear that satellites would have military utility, especially for reconnaissance. As early as 1947, the Soviet press denounced the prospect of U.S. satellites as "instruments of blackmail." Obviously, the idea of U.S. cameras orbiting with impunity was extremely unnerving to the secretive Soviet state, and an October 4, 1950, RAND Corporation report warned that the response of the Soviet Union to the

---

* Asteroid 2002 AA29, a 60-meter object discovered in 2002, is also in an Earth-synchronized solar orbit.

launching of a U.S. satellite might be dangerous. The RAND report concluded that any such launch should be done with advance publicity, especially if the spacecraft was being stressed as a nonweapon.

It was assumed that the Soviets would regard satellite reconnaissance as an attack on their secrecy and their sovereignty, and therefore would consider it to be illegal. But was it illegal? It was an open question. Overflight of a nonassenting nation was contrary to international law, but was there an upper limit to the airspace in question? According to historian Walter A. McDougall, "It was very doubtful that the USSR would accept any vertical limitation on its sovereignty or accept that any passage of a spacecraft over its territory might be innocent. Rather, orbiting a satellite over the Soviet Union might be construed by the Kremlin as an act of aggression."[6] McDougall further points out that "just as important as developing such [satellite reconnaissance] technology was establishing the legal right to use it." The RAND report concluded, "Our objective is to reduce the effectiveness of any Soviet counteraction. . . . Perhaps the best way to minimize the risk of countermeasures would be to launch an 'experimental' satellite on an equatorial orbit." Such a satellite would not have an overt military mission and would not overfly Soviet territory, so it could test the "freedom of space" issue in the best political environment.[7]

By the mid-1950s, the Eisenhower administration was studying the possibility of a nuclear test ban treaty and the freezing or limiting of the deployment of nuclear weapons. The main problem of an arms control agreement was verification. Also, in the absence of such an agreement, management of an arms race required adequate intelligence of Soviet capabilities. However, Soviet leader Nikita Khrushchev refused to allow on-site inspections and rejected President Eisenhower's Open Skies proposal, which would permit overflights of each other's national territory. The United States had a greater need to conduct reconnaissance on the closed society of the Soviet Union than the Soviet Union had to spy on the open society of the United States. In June 1956, U-2 high altitude reconnaissance aircraft began secret, illegal overflights of the Soviet Union. Meanwhile the U.S. Air Force continued WS-117L, a program that began in March 1955 to develop a "strategic satellite system."

Against this backdrop of intelligence requirements, the Eisenhower administration returned to the question of how to establish the legal precedent of the "freedom of space." The opportunity for a credible, innocuous satellite program presented itself on October 4, 1954, when the Special Committee for the International Geophysical Year recommended that participating governments launch satellites in the interest of science.[8] In 1954, the United States already had a missile powerful enough to serve as the basis for a satellite launch vehicle. The Redstone missile simply needed a cluster of small, "off the shelf," solid-fuel rockets to serve as upper stages. The original purpose and heritage of the missile, however, complicated matters. The Redstone missile was an intermediate range ballistic missile (IRBM), designed by the U.S. Army to deliver a nuclear warhead. Furthermore, it was the product of a team of German engineers headed by Wernher von Braun, who had developed the A-4 (popularized as the V-2 by Joseph Goebbels's propaganda ministry). The A-4 was the world's first ballistic missile weapon, developed during World War II, and the Redstone missile was a modest outgrowth of the A-4. In addition to the Redstone missile, the United States had the Viking sounding rocket, which the U.S. Naval Research Laboratory had developed for scientific purposes. The Navy's Viking was much less powerful than the Army's Redstone, and the upper stages necessary for it to put a satellite in orbit would have to be designed from scratch, whereas von Braun's design for a space launch vehicle was based on existing hardware. The Army configuration could do the job in a few months; the Navy configuration (later called Vanguard) hoped to achieve a launch within four years, by the end of 1958. On August 3, 1955, the Stewart Committee voted 3 to 2 in favor of the Navy's proposal for Viking.[9]

In 1955, both the United States and the Soviet Union announced their intention to launch satellites into orbit around the Earth as part of their scientific investigations during the International Geophysical Year (IGY). No state objected. Thus, the right of orbital overflight, recognized from the very beginning by the two original launching states and not objected to by any other states, became a customary norm virtually instantaneously. This norm continued to hold as other states placed satellites in orbit, either by their own means or by the means of other launching states. However, a more precise delimitation of space continues to be a problem of international law.[10]

In retrospect, the Stewart Committee's split decision in favor of the Navy proposal may look like a quirk of history. Had the decision gone the other way, the United States could have launched a satellite more than a year before *Sputnik 1*. However, the decision was not at all out of step with national policy. There was a critical need to monitor Soviet development of intercontinental ballistic missiles (ICBM), so the first priority was to establish the legality of satellite overflight to clear the path for reconnaissance spacecraft. There were two ways to do this. One way was for the United States to launch a small satellite for the advancement of science, with no one objecting to it. The other way was for the Soviet Union to launch the first satellite. The second solution was less desirable because of the prestige gained from launching the first satellite, "but it was not worth taking every measure to prevent."[11]

Another reason for treating outer space as a commons was to avoid claims of national sovereignty, and thereby avoid the need for nation-states to compete not only to stake such claims, but to create the capacity to defend such claims. Neither the Soviet Union nor the United States desired the militarization of outer space, a costly race to secure territories of no immediate military or economic utility, and of only imagined future value.

M. J. Peterson, professor of political science at the University of Massachusetts, confirms that the convergence on the conception of outer space as a commons occurred quite early:

> [A] realist would not have been able to predict whether outer space would be treated as a common area or as something to be "conquered" and parceled out among spacefaring states. Both conceptions of space were advanced in the early 1950s; some commentators compared space to the high seas, while others compared it to national airspace. . . . [T]he superpowers initially disagreed, with the U.S. government preferring the high seas conception and the Soviet government the national airspace conception.
>
> The development of outer space law did involve moments of compromise, but the decision to treat space as a commons involved a clear choice of one conception over the other, an outcome that depended on the Soviet government's shift to accepting the high seas conception. The process by which convergence occurred can be traced in some detail because outer space law was developed in a well-documented multilateral negotiation.[12]

Writing as the international community was coming to grips with the problem of sovereignty over the Moon and other celestial bodies, Modesto Seara Vázquez of National University of Mexico struggles with the opposing concepts of *res nullius* (something that belongs to none) and *res communis* (something that belongs to all), the former being appropriable by the first occupant, the latter not being appropriable.

> Today they belong to no one, but are they *res nullius?* If they are, they will belong to the first to occupy them, as long as the occupation fulfills the required conditions. But we doubt that the nations would accept the first occupant as the sole sovereign of a celestial body.[13]

Seara Vázquez concludes that the concept of *res nullius* is disappearing from international law, and that it will continue to apply to celestial bodies only while they remain beyond the reach of humankind.

> But the moment the possibility of reaching them appears, they are no longer considered *res nullius*. We think, therefore, that they can only be accepted as *res nullius* provisionally, that is, as long as they are out of man's reach. Afterwards, they become something else.[14]

Is that "something else" *res communis?* Seara Vázquez lists the characteristics of *res communis*, both positively and negatively. Positively:

- They are things of nature.
- They are for the common use of all men.
- Their use is regulated by national or international law.

Negatively:

- They are not susceptible to becoming objects of the right of property.
- No nation, corporation, or individual, may appropriate them.[15]

The first *res communis* to emerge in international law was the sea, per Hugo Grotius's theory of the freedom of the seas, which included both freedom of navigation and freedom to extract natural resources.[16] By extension, the airspace above international waters became *res communis*, pursuant to the United Nation's 1944 Convention on International Civil Aviation, and thus open to all aviation.[17] Finally, in the UN's 1959 Antarctic Treaty, an entire continent became *res communis*.[18]* Thus, each of the three distinct environments of the Earth—sea, air, and land—contain examples of *res communis*. The specific case of fishing rights in international waters (outside of exclusive economic zones) demonstrates the principle that *res communis* means that all have an equal right to exploit the natural resources of the commons, but not a right to share equally in what is exploited. The right to fish in international waters was unrestricted as long as this natural resource was considered unlimited; eventually, however, the need arose for an international regime to provide for the rational management of what came to be viewed as a limited resource. This principle was reflected in the 1958 Convention on Fishing and Conservation of Living Resources of the High Seas, one of the four treaties negotiated under the first United Nations Conference on Law of the Sea (UNCLOS I).[19]

Seara Vázquez predicts that that the same process will unfold on the celestial bodies:

> According to this, what will happen is that a regime of co-sovereignty will be formed, through a moral entity, which will be the only practical solution acceptable.

---

* The initial status of Antarctica under the treaty was a bit more complicated than this. States had already laid claim to portions of the continent, and a number of these claims overlapped. Parties to the treaty agreed to hold their claims in abeyance for the duration of the treaty. To date, no state has renewed its claim.

In fact, as soon as technological progress makes exploitation of celestial bodies possible, all the states will want to take the fullest advantage, and there will inevitably be conflicts.

In summing up the legal status of celestial bodies, it can be concluded: in principle they are *res nullius,* since they belong to no one and any state may appropriate them. By the practice being followed, they belong rather to *res communis,* because it is the will of all that there be no exclusive appropriation for the benefit of one special country or group of countries. They prefer exploitation in common. But celestial bodies are not *res communis* in the proper sense of the word, because this means use by all, without sovereignty on the part of any.[20]

All of this creates the unsettling feeling that neither *res nullius* nor *res communis* adequately describes the legal character of the celestial bodies, that there is something about the cosmic commons that is quite uncommon.

According to Seara Vázquez a private party could not acquire any rights over territory, nor would it be able to occupy any territory in the name of a state. Whatever rights it had under the public law of a state on Earth would not exist on the celestial body. In the 1960s, at the time Seara Vázquez writes, it was already clear that states would not claim the right to occupy celestial bodies under the principle of *res nullius.* Therefore, he arrives at his third hypothesis: that the community of nations would occupy celestial bodies. This conclusion in turn spawns four possibilities:

1. Occupation is open to all states capable of reaching a celestial body; however, this maintains a situation of privilege for these states.
2. The celestial body is divided into reserved zones of occupation; however, this presents the problem of the distribution of the zones among states, as well as the problem of inefficient utilization of resources.
3. Occupation is vested in indivisible co-sovereignty, which gives each state the right to use the celestial body as suits its interest; however, this is only acquiescence to anarchy, leading to the eventual triumph of the strong over the weak.
4. Occupation is vested in a moral entity that represents all nations.

Seara Vázquez looks to this last option as being the most feasible:

It is very possible that the commissions created by the United Nations for cosmic space will develop to the point of becoming the very organization we propose, as they go on accumulating competency on different political or legal aspects of the exploration and use of space.[21]

Jenks suggests that

[i]t would seem desirable to start from the principle that title to the natural resources of the Moon and of other planets and satellites should be regarded as vested in the United Nations and that any exploitation of such resources which might be possible should be on the basis of concessions, leases or licenses from the United Nations.[22]

The development of the international law of outer space is the story of the evolution of the moral entity suggested by Seara Vázquez, in a process by which disparate national interests, economic systems, and legal traditions contend to hammer out a common vision of celestial governance. At this point, it will be useful to briefly review the international declarations, resolutions, treaties, and agreements through which the legal characterization of outer space and the celestial bodies has developed, along with the legal principles to guide the activities of states.

APPLICABLE TREATIES, DECLARATIONS, AND RESOLUTIONS

Tables 2.1 through 2.3 provide a summary of the international and transnational documents dealing with issues of national sovereignty and private property rights in outer space. The United Nations resolutions and declarations in Table 2.1 provided the first unified expression of the principles upon which later, legally binding, international documents

Table 2.1: Applicable United Nations Resolutions and Declarations

Full Name	Short Name	Number	Date
Resolution on International Cooperation in the Peaceful Uses of Outer Space	International Cooperation Resolution	Res 1721 (XVI)	December 20, 1961
Declaration of Legal Principles Governing the Activities of States in the Exploration and Uses of Outer Space	Declaration of Legal Principles	Res 1962 (XVIII)	December 13, 1963
Declaration on International Cooperation in the Exploration and Use of Outer Space for the Benefit and in the Interest of All States, Taking into Particular Account the Needs of Developing Countries	Declaration on International Cooperation	A/Res 51/122	December 13, 1996

Table 2.2: Applicable United Nations Treaties

Full Name	Short Name	Opened	Entered Force	As of January 1, 2008 Ratifications	Signatures
Treaty on Principles Governing the Activities of States in the Exploration and Use of Outer Space, including the Moon and Other Celestial Bodies	Outer Space Treaty	January 27, 1967	October 10, 1967	98	27
Convention on Registration of Objects Launched into Outer Space	Registration Convention	January 14, 1975	September 15, 1976	51	4
Agreement Governing the Activities of States on the Moon and Other Celestial Bodies	Moon Agreement	December 18, 1979	July 11, 1984	13	4

would be based. Table 2.2 shows the three international treaties that are currently in force. These treaties have substantially different stature in international law: two have been nearly universally embraced, whereas the other has been nearly universally shunned. The documents in Table 2.3 are not generally recognized sources of international law; rather they are objects of debate.

The International Cooperation Resolution

United Nations General Assembly Resolution 1721 (XVI) was adopted without vote on December 20, 1961 (see appendix 1).[23] Commonly called the International Cooperation Resolution, it established the principle of national non-appropriation upon which would be concluded later international agreements having the force of law. The language of this legacy

Table 2.3: Applicable Multilateral Declarations and Proposed Documents

Full Name	Short Name	Originator	Type	Date
Declaration of the First Meeting of Equatorial Countries	Bogotá Declaration	Equatorial Countries	Multilateral	December 3, 1976
Act to Promote Privately Funded Space Exploration and Settlement	Space Settlement Prize Act	Space Settlement Institute	National	1997
Convention on Jurisdiction and Real Property Rights in Outer Space	Property Rights Convention	Wayne N. White	International	2001
Constitution for the Regency of United Societies in Space	ROUSIS Constitution	United Societies in Space	Transnational	August 4, 2001

Sources: Colombia, Ecuador, Brazil, Congo, Zaire, Uganda, Kenya, and Indonesia. 1976. "Declaration of the First Meeting of Equatorial Countries." Available from http://www. jaxa.jp/jda/library/space-law/chapter_2/2-2-1-2_e.html; accessed: October 28, 2004; Wasser, Alan. 2004. "The Space Settlement Prize Act." Available from http://www. spacesettlement.org/law/; accessed January 28, 2005; White, Wayne N. 2001. "Proposal for a Multilateral Treaty Regarding Jurisdiction and Real Property Rights in Outer Space." Internet. Available from http://www.spacefuture.com/archive/proposal_for_a_multilateral_ treaty_regarding_jurisdiction_and_real_property_rights_in_outer_space.shtml; accessed December 26, 2005; and O'Donnell, Declan J. 2001. "Constitution for the Regency of United Societies in Space." *Space Governance Journal,* 6, 22. Available from http://www.angelfire. com/space/usis/constitution.html; accessed February 22, 2006.

document and UN General Assembly Resolution 1962 (XVIII), commonly called the Declaration of Legal Principles, also refutes claims made by some space property rights advocates regarding the political motivations of the Johnson administration in concluding the 1967 Outer Space Treaty.

The Declaration of Legal Principles

United Nations General Assembly Resolution 1962 (XVIII) was adopted without vote on December 13, 1963 (see Appendix 2).[24] Although the Declaration of Legal Principles is not a treaty, Hungarian barrister Imre Csabafi declared that because the space powers agreed on the UN General Assembly as the forum for this instrument, and the instrument was universally adopted, "it cannot be denied then, that the norms declared by Resolution 1962 (XVIII) . . . are binding rules of space law."[25] The form of the declaration (i.e., a UN General Assembly resolution) was an interim step toward the 1967 Outer Space Treaty.

It is worth noting that a provision in the Soviet Union's 1962 Draft Declaration of the Basic Principles Governing the Activities of States Pertaining to the Exploration and Uses of Outer Space can be construed as an attempt to keep capitalism grounded:

> All activities of any kind pertaining to the exploration of outer space shall be carried out solely and exclusively by States.[26]

The attempt was not successful. The United States passed the Communications Satellite Act in 1962, establishing the right of private entities to conduct space operations and laying the groundwork for the formation of the Communications Satellite Corporation (COMSAT).[27] The Soviets withdrew their proposal of government-only operations in space the following year, declaring, however, that

> The Soviet delegation considers it essential to point out that in this field it would be possible to consider the question of not excluding from the declaration the possibility of activity in outer space by private companies, on the condition that such activity would be subject to the control of the appropriate State, and the State would bear international responsibility for it.[28]

The point was well taken: states are the subjects of international law, whereas natural and juridical persons generally are not. International law regulates the activities of states; in turn, municipal law regulates the activities of natural and juridical persons within their jurisdiction. This principle was adopted in the 1963 Declaration of Legal Principles, which defined "national activities" as those "carried on by governmental agencies or non-governmental entities." The principle was later reflected in the 1967 Outer Space Treaty.

The Outer Space Treaty

The Treaty on Principles Governing the Activities of States in the Exploration and Use of Outer Space, including the Moon and Other Celestial Bodies, commonly referred to as the Outer Space Treaty, codified the legal principles set forth in the 1963 resolution-declaration (see appendix 3).[29] The treaty was opened for signature on January 27, 1967, and was immediately signed by the Soviet Union, the United Kingdom, and the United States. The treaty entered force on October 10, 1967. As of January 2003, 98 states had ratified the treaty, including all launching states (see Table 2.4), and another 27 states had signed it. Of the 30 nonlaunching states that have had their satellites placed in orbit by others, only six are not party to the treaty, and they account for less than half percent of all the spacecraft that have been placed in orbit. Thus, 99.6 percent of all satellites have been placed in orbit by states.

Table 2.4: Launching States

Launching State	First Satellite		Outer Space Treaty	Moon Agreement
	Name	Launch Date		
USSR/Russia	*Sputnik 1*	October 4, 1957	Ratified October 10, 1967	
USA	*Explorer 1*	February 1, 1958	Ratified October 10, 1967	
France/ESA	*Asterix 1*	November 26, 1965	Ratified August 5, 1970	Signed January 29, 1980
Japan	*Osumi*	February 11, 1970	Ratified October 10, 1967	
China	*Dong Fang Hong 1*	April 24, 1970	Acceded December 20, 1983	
UK	*Prospero*	October 28, 1971	Ratified October 10, 1967	
India	*Rohini 1B*	July 18, 1980	Ratified January 18, 1982	Signed January 18, 1982
Israel	*Offek 1*	September 19, 1988	Ratified February 18, 1977	
Iran	*Omid*	August 17, 2008	Signed January 27, 1967	

Sources: www.tbs-satellite.com/tse/online/thema_first.html and www.state.gov/t/ac/trt/5181.htm

The Registration Convention

Although the Convention on Registration of Objects Launched into Outer Space, the Registration Convention, contains no mention of sovereignty or property rights, it does contain language regarding space objects and orbits (see appendix 4).[30] It is one of the international legal documents that provides the tie between the two and is an example of the functional definition of outer space.

The Moon Agreement

The Agreement Governing the Activities of States on the Moon and Other Celestial Bodies, the Moon Agreement, was negotiated during the same years as the third United Nations Convention on the Law of the Sea (UNCLOS III).[31] The principles of the international law of outer space in large part derive from the international law of the sea, and both environments are regarded as commons, so it is not surprising that the two treaties contain similar provisions regarding the use of natural resources. The Moon Agreement was opened for signature on December 18, 1979 (see appendix 6). The Moon Agreement technically entered force on July 11, 1984, 30 days after ratification by the fifth state. As of January 2003, only 13 states had ratified the agreement, and another four states had signed it. Although most of its provisions are not binding on any launching states, it may be viewed as a source of legal principles. Furthermore, P. P. C. Haanappel, associate professor of law at McGill University, states:

> Certain paragraphs of Article XI are drafted as being declaratory of existing international law, thus binding on all States. Others are drafted in a fashion imposing obligations on Contracting Parties only. Of the former kind seem to be paragraphs 1 to 3 and probably 8.[32]

If the paragraphs Haanappel cites are indeed compelling law binding on all states, or *jus cogens,* it should be noted that paragraphs 1 and 3 refer to paragraph 5 of the same article, and paragraph 8 refers to paragraph 7 of the same article. Both paragraphs 5 and 7 address the specifics of establishing a governing regime for the exploitation of lunar resources, and do not express any principles, thus they cannot be considered *jus cogens.* Paragraph 8 also refers to Article 6, paragraph 2, which does declare principles; thus, these provisions of the agreement, along with the declaratory portions of paragraph 1 through 3, may be considered *jus cogens.*

The Declaration on International Cooperation

The shift to a market economy in China in the 1980s, the overthrow of communist governments in eastern Europe in 1989, and the breakup of the Soviet Union in 1991 all affected the Declaration on International Cooperation in the Exploration and Use of Outer Space for the Benefit and in the Interest of All States, Taking into Particular Account the Needs of Developing Countries.[33] The Declaration on International Cooperation reflects a watering down of principles in earlier documents regarding the sharing of benefits (see Appendix 9).

LEGAL PRINCIPLES

The Cosmic Commons

The Declaration of Legal Principles

Principle 1
The exploration and use of outer space shall be carried on for the benefit and in the interests of all mankind.

Principle 2
Outer space and celestial bodies are free for exploration and use by all States on a basis of equality and in accordance with international law.

The Outer Space Treaty

Article 1 elaborates on Principles 1 and 2 of the 1963 Declaration of Legal Principles of Outer Space to include the interests of nonlaunching states:

> The exploration and use of outer space, including the moon and other celestial bodies, shall be carried out for the benefit and in the interests of all countries, irrespective of their degree of economic or scientific development, and shall be the province of all mankind.
>
> Outer space, including the moon and other celestial bodies, shall be free for exploration and use by all States without discrimination of any kind, on a basis of equality and in accordance with international law, and there shall be free access to all areas of celestial bodies.
>
> There shall be freedom of scientific investigation in outer space, including the moon and other celestial bodies, and States shall facilitate and encourage international co-operation in such investigation.

The Moon Agreement

The language of Article 4, paragraph 1, is generally in line with the sentiment of Principle 1 of the 1963 Declaration of Legal Principles of Outer Space and Article 1, paragraph 1, of the 1967 Outer Space Treaty, but is more specific on the desirability for the benefits of the exploration and use of the moon to be shared with developing countries and future generations:

> The exploration and use of the moon shall be the province of all mankind and shall be carried out for the benefit and in the interests of all countries, irrespective of their degree of economic or scientific development. Due regard shall be paid to interests of present and future generations as well as to the need to promote higher standards of living conditions of economic and social progress and development in accordance with the Charter of the United Nations.

Article 6, Paragraph 1
> There shall be freedom of scientific investigation on the moon by all States Parties without discrimination of any kind, on the basis of equality and in accordance with international law.

Article 11, Paragraph 1
> The moon and its natural resources are the common heritage of mankind, which finds its expression in the provisions of this Agreement, in particular in paragraph 5 of this article.

(Cited by Haanappel as binding on all.)[34]

Article 11, Paragraph 4
> States Parties have the right to exploration and use of the moon without discrimination of any kind, on the basis of equality and in accordance with international law and the provisions of this Agreement.

The Declaration on International Cooperation

Paragraph 2
> States are free to determine all aspects of their participation in international cooperation in the exploration and use of outer space on an equitable and mutually acceptable basis. Contractual terms in such cooperative ventures should be fair and reasonable and they should be in full compliance with the legitimate rights and interests of the parties concerned as, for example, with intellectual property rights.

National Appropriation

The International Cooperation Resolution

Paragraph A(1)(b)
> Outer space and celestial bodies are free for exploration and use by all States in conformity with international law and are not subject to national appropriation.

The Declaration of Legal Principles

Principle 3
Outer space and celestial bodies are not subject to national appropriation by claim of sovereignty, by means of use or occupation, or by any other means.

The Outer Space Treaty

Article 2
Outer space, including the moon and other celestial bodies, is not subject to national appropriation by claim of sovereignty, by means of use or occupation, or by any other means.

The Moon Agreement

Article 11, Paragraph 2
The moon is not subject to national appropriation by any claim of sovereignty, by means of use or occupation, or by any other means.

(Cited by Haanappel as binding on all.)[35]

International Responsibility

The Declaration of Legal Principles

Principle 5
States bear international responsibility for national activities in outer space, whether carried on by governmental agencies or by non-governmental entities, and for assuring that national activities are carried on in conformity with the principles set forth in the present Declaration. The activities of non-governmental entities in outer space shall require authorization and continuing supervision by the State concerned. When activities are carried on in outer space by an international organization, responsibility for compliance with the principles set forth in this Declaration shall be borne by the international organization and by the States participating in it.

The Outer Space Treaty

Article 6
States Parties to the Treaty shall bear international responsibility for national activities in outer space, including the moon and other celestial bodies, whether such activities are carried on by governmental agencies or by non-governmental entities, and for assuring that national activities are carried out in conformity with the provisions set forth in the present Treaty.

The activities of non-governmental entities in outer space, including the moon and other celestial bodies, shall require authorization and continuing supervision by the appropriate State Party to the Treaty. When activities are carried on in outer space, including the moon and other celestial bodies, by an international organization, responsibility for compliance with this Treaty shall be borne both by the international organization and by the States Parties to the Treaty participating in such organization.

The Moon Agreement

Article 14, Paragraph 1

States Parties to this Agreement shall bear international responsibility for national activities on the moon, whether such activities are carried out by governmental agencies or by non-governmental entities, and for assuring that national activities are carried out in conformity with the provisions of this Agreement. States Parties shall ensure that non-governmental entities under their jurisdiction shall engage in activities on the moon only under the authority and continuing supervision of the appropriate State Party.

Freedom from Interference

The Declaration of Legal Principles

Principle 6

In the exploration and use of outer space, States shall be guided by the principle of co-operation and mutual assistance and shall conduct all their activities in outer space with due regard for the corresponding interests of other States. If a State has reason to believe that an outer space activity or experiment planned by it or its nationals would cause potentially harmful interference with activities of other States in the peaceful exploration and use of outer space, it shall undertake appropriate international consultations before proceeding with any such activity or experiment. A State which has reason to believe that an outer space activity or experiment planned by another State would cause potentially harmful interference with activities in the peaceful exploration and use of outer space may request consultation concerning the activity or experiment.

The Outer Space Treaty

Article 9

In the exploration and use of outer space, including the Moon and other celestial bodies, States Parties to the Treaty shall be guided by the principle of co-operation and mutual assistance and shall conduct all their activities in outer space, including the Moon and other celestial bodies, with due regard to the corresponding interests of all other States Parties to the Treaty. States Parties to the Treaty shall pursue studies of outer space, including the Moon and other celestial bodies, and conduct exploration of them so as to avoid their harmful contamination and also adverse changes in the environment of the Earth resulting from the introduction of extra-terrestrial matter and, where necessary, shall adopt appropriate measures for this purpose. If a State Party to the Treaty has reason to believe that an activity or experiment planned by it or its nationals in outer space, including the Moon and other celestial bodies, would cause potentially harmful interference with activities of other States Parties in the peaceful exploration and use of outer space, including the Moon and other celestial bodies, it shall undertake appropriate international consultations before proceeding with any such activity or experiment. A State Party to the Treaty which has reason to believe that an activity or experiment planned by another State Party in outer space, including the Moon and other celestial bodies, would cause potentially harmful interference with activities in the peaceful exploration and use of outer space, including the Moon and other celestial bodies, may request consultation concerning the activity or experiment.

The Moon Agreement

Article 5, Paragraph 2

If a State Party becomes aware that another State Party plans to operate simultaneously in the same area of or in the same orbit around or trajectory to or around the moon, it shall promptly inform the other State of the timing of and plans for its own operations.

Article 8, Paragraph 3

Activities of States Parties in accordance with paragraphs 1 and 2 of this article shall not interfere with the activities of other States Parties on the moon. Where such interference may occur, the States Parties concerned shall undertake consultations in accordance with article 15, paragraphs 2 and 3, of this Agreement.

Jurisdiction over and Ownership of Space Objects

The Declaration of Legal Principles

Principle 7

The State on whose registry an object launched into outer space is carried shall retain jurisdiction and control over such object, and any personnel thereon, while in outer space. Ownership of objects launched into outer space, and of their component parts, is not affected by their passage through outer space or by their return to the earth. Such objects or component parts found beyond the limits of the State of registry shall be returned to that State, which shall furnish identifying data upon request prior to return.

The Outer Space Treaty

Article 8

A State Party to the Treaty on whose registry an object launched into outer space is carried shall retain jurisdiction and control over such object, and over any personnel thereof, while in outer space or on a celestial body. Ownership of objects launched into outer space, including objects landed or constructed on a celestial body, and of their component parts, is not affected by their presence in outer space or on a celestial body or by their return to the Earth. Such objects or component parts found beyond the limits of the State Party to the Treaty on whose registry they are carried shall be returned to that State Party, which shall, upon request, furnish identifying data prior to their return.

The Moon Agreement

Article 11, Paragraph 3

Neither the surface nor the subsurface of the moon, nor any part thereof or natural resources in place, shall become property of any State, international intergovernmental or non-governmental organization, national organization or non-governmental entity or of any natural person. The placement of personnel, space vehicles, equipment, facilities, stations and installations on or below the surface of the moon, including structures connected with its surface or subsurface, shall not create a right of ownership over the surface or the subsurface of the moon or any areas thereof. The foregoing provisions are without prejudice to the international regime referred to in paragraph 5 of this article.

(Cited by Haanappel as binding on all.)[36]

Article 12, Paragraph 1

States Parties shall retain jurisdiction and control over their personnel, space vehicles, equipment, facilities, stations and installations on the moon. The ownership of space vehicles, equipment, facilities, stations and installations shall not be affected by their presence on the moon.

Visitation

The Outer Space Treaty

Article 7

All stations, installations, equipment and space vehicles on the Moon and other celestial bodies shall be open to representatives of other States

Parties to the Treaty on a basis of reciprocity. Such representatives shall give reasonable advance notice of a projected visit, in order that appropriate consultations may be held and that maximum precautions may be taken to assure safety and to avoid interference with normal operations in the facility to be visited.

The Moon Agreement

Article 15, Paragraph 1

Each State Party may assure itself that the activities of other States Parties in the exploration and use of the moon are compatible with the provisions of this Agreement. To this end, all space vehicles, equipment, facilities, stations and installations on the moon shall be open to other States Parties. Such States Parties shall give reasonable advance notice of a projected visit, in order that appropriate consultations may be held and that maximum precautions may be taken to assure safety and to avoid interference with normal operations in the facility to be visited. In pursuance of this article, any State Party may act on its own behalf or with the full or partial assistance of any other State Party or through appropriate international procedures within the framework of the United Nations and in accordance with the Charter.

Resource Extraction

The Moon Agreement

Article 6, Paragraph 2

In carrying out scientific investigations and in furtherance of the provisions of this Agreement, the States Parties shall have the right to collect on and remove from the moon samples of its mineral and other substances. Such samples shall remain at the disposal of those States Parties which caused them to be collected and may be used by them for scientific purposes. States Parties shall have regard to the desirability of making a portion of such samples available to other interested States Parties and the international scientific community for scientific investigation. States Parties may in the course of scientific investigations also use mineral and other substances of the moon in quantities appropriate for the support of their missions.

(Cited by Haanappel as probably binding on all by virtue of reference by Article 11, paragraph 8.)[37]

Article 11, Paragraph 5

States Parties to this Agreement hereby undertake to establish an international regime, including appropriate procedures, to govern the exploitation of the natural resources of the moon as such exploitation is about to become feasible. This provision shall be implemented in accordance with article 18 of this Agreement.

Article 11, Paragraph 7

The main purposes of the international regime to be established shall include:

1. The orderly and safe development of the natural resources of the moon;
2. The rational management of those resources;
3. The expansion of opportunities in the use of those resources;
4. An equitable sharing by all States Parties in the benefits derived from those resources, whereby the interests and needs of the developing countries, as well as the efforts of those countries which have contributed either directly or indirectly to the exploration of the moon, shall be given special consideration.

Use Restriction

The Moon Agreement

Article 9, Paragraph 1

States Parties may establish manned and unmanned stations on the moon. A State Party establishing a station shall use only that area which is required for the needs of the station and shall immediately inform the Secretary-General of the United Nations of the location and purposes of that station. Subsequently, at annual intervals that State shall likewise inform the Secretary-General whether the station continues in use and whether its purposes have changed.

Article 9, Paragraph 2

Stations shall be installed in such a manner that they do not impede the free access to all areas of the moon of personnel, vehicles and equipment of other States Parties conducting activities on the moon in accordance with the provisions of this Agreement or of article I of the Treaty of Principles Governing the Activities of States in the Exploration and Use of Outer Space, including the Moon and other Celestial Bodies.

NOTES

1. Locke, John 1689. *The True Original, Extent, and End of Civil-Government.* Available from http://www.constitution.org/jl/2ndtr00.htm; accessed December 26, 2005.

2. Hobbes, Thomas. 1651. *Leviathan.* Available from http://oregonstate.edu/instruct/phl302/texts/hobbes/leviathan-contents.html; accessed December 28, 2005.

3. Haley, A. G. 1956. "Basic Concepts of Space Law." *Jet Propulsion,* 26: 951–957.

4. Cooper, John C. 1951. "High Altitude Flight and National Sovereignty." *International Law Quarterly,* 4: 411.

5. Schachter, Oscar. 1952. "Legal Aspects of Space Travel." *Journal of the British Interplanetary Society*, 11, 1:3–6; Meyer, A. 1952. "Legal Problems of Flight Into the Outer Space." *Annual Report of the British Interplanetary Society*, 353–254; Jenks, C. Wilfred. 1956 (January). "International Law and Activities in Space." *International and Comparative Law Quarterly*, 99–114.

6. McDougall, Walter A. 1985. *The Heavens and the Earth: A Political History of the Space Age.* New York: Basic Books, 107–109.

7. Ibid., 109–110.

8. Ibid., 114–118.

9. Ibid., 119–122.

10. Gangale, Thomas. 2006. "Who Owns the Geostationary Orbit?" *Annals of Air and Space Law*, 31:425–446. Montreal, Quebec.

11. McDougall, *Heavens and Earth*, 122–124

12. Peterson, M. J. 1997. "The Use of Analogies in Developing Outer Space Law." *International Organization*, 51, 2:245–274.

13. Seara Vázquez, Modesto. 1965. *Cosmic International Law*. Detroit: Wayne State University Press, 116.

14. Ibid., 117.

15. Ibid., 216

16. Grotius, Hugo. 1609. *Mare Liberum*. Available from http://socserv2.socsci.mcmaster.ca/~econ/ugcm/3ll3/grotius/Seas.pdf; accessed January 25, 2006.

17. United Nations. 1944. "Convention on International Civil Aviation." 15 U.N.T.S. 295. Available from http://www.iasl.mcgill.ca/airlaw/public/chicago/chicago1944a.pdf; accessed November 29, 2004.

18. United Nations. 1959. "Antarctic Treaty." 402 U.N.T.S. 71. Available from http://www.nsf.gov/od/opp/antarct/anttrty.htm; accessed October 31, 2004.

19. United Nations. 1958. "Convention on Fishing and Conservation of Living Resources of the High Seas." 559 U.N.T.S. 285. Available from http://www.un.org/law/ilc/texts/fishfra.htm; accessed January 11, 2006.

20. Seara Vázquez, *Cosmic International Law*, 221–222

21. Ibid.

22. Jenks, C. Wilfred. 1962. *The Common Law of Mankind*. New York: Praeger, 398.

23. United Nations. 1961 (December 20). "International Cooperation in the Peaceful Uses of Outer Space." General Assembly Resolution 1721 (XVI).

24. United Nations. 1963 (December 13). "Declaration of Legal Principles Governing the Activities of States in the Exploration and Uses of Outer Space." General Assembly Resolution 1962 (XVIII).

25. Csabafi, Imre. 1965. "The UN General Assembly Resolutions on Outer Space as Sources of Space Law." *Proceedings, 8th Colloquium on the Law of Outer Space.* American Institute of Aeronautics and Astronautics, 337–361.

26. United Nations Committee on the Peaceful Uses of Outer Space. 1962. U.N. Doc. A/AC.105/L.2.

27. United States Congress. 1962. "Communications Satellite Act of 1962." Available from http://www.presageinc.com/contents/experience/satellitereform/contents/briefingbook/technology/1962act.pdf; accessed October 23, 2008.

28. United Nations Committee on the Peaceful Uses of Outer Space. 1963. U.N. Doc. A/AC.105/PV.22, 37.

29. United Nations. 1967. "Treaty on Principles Governing the Activities of States in the Exploration of Outer Space, Including the Moon and other Celestial Bodies." 610 U.N.T.S. 205. Available from http://www.iasl.mcgill.ca/spacelaw/outerspace.html; accessed September 17, 2004.

30. United Nations. 1974. "Convention on Registration of Objects Launched Into Outer Space." 1023 U.N.T.S. 15. Available from http://www.iasl.mcgill.ca/spacelaw/registration.html; accessed September 17, 2004.

31. United Nations. 1979. "Agreement Governing the Activities of States on the Moon and Other Celestial Bodies." 1363 U.N.T.S. 3. Available from http://www.iasl.mcgill.ca/spacelaw/moon.html; accessed September 17, 2004; and United Nations. 1982. "United Nations Convention on the Law of the Sea." 1833 U.N.T.S. 3. Available from http://www.un.org/Depts/los/convention_agreements/texts/unclos/unclos_e.pdf; accessed November 29, 2004.

32. Haanappel, P. P. C. 1980. "Article XI of the Moon Treaty." *Proceedings, 23rd Colloquium on the Law of Outer Space.* American Institute of Aeronautics and Astronautics, 80-SL-10, 29–33.

33. United Nations. 1996. "Declaration on International Cooperation in the Exploration and Use of Outer Space for the Benefit and in the Interest of all States, Taking into Particular Account the Needs of Developing Countries." G.A. Res. 51/122. Internet. Available from http://www.unoosa.org/oosa/SpaceLaw/spben.html; accessed February 22, 2006.

34. Haanappel, "Article XI of the Moon Treaty," 29–33.

35. Ibid.

36. Ibid.

37. Ibid.

3

Property Rights in Outer Space

Do property rights exist in outer space under the current international legal framework? If so, to what sort of property do they apply and under what circumstances? What property rights are necessary or desirable that may not be protected under current law? Does the Outer Space Treaty undercut or provide the basis in principle for property rights? Can specific protections be secured without the necessity of negotiating new international treaties?

NONGOVERNMENTAL APPROPRIATION OF REAL PROPERTY

With the purpose of finding a private real property right on celestial bodies, Wayne N. White, author on space law, expostulates on the differences between common law and civil law regarding property rights:

> The relationship between property and sovereignty differs under common law and civil law systems. The common law theory of title has its roots in feudal law. Under this theory the Crown holds the ultimate title to all lands, and the proprietary rights of the subject are explained in terms of vassalage. Civil law, on the other hand, is derived from Roman law, which distinguishes between property and sovereignty. Under this theory it is possible for property to exist in the absence of sovereignty.[1]

This seems to overstate things. It may be that "under this theory it is possible for property to exist in the absence of sovereignty," just as it is possible for a planet to exist in the absence of its discovery, or for gravity to exist as a force of nature in the absence of a scientific theory. What practical use can we make of such an argument? Practically speaking, property can only exist if there is a recognized right to the property. That recognition must originate in some legal regime, which in turn implies either an act of national sovereignty or, as in the case of an international regime, of collective sovereignty. White continues:

> Article 2 of the Outer Space Treaty prohibits territorial sovereignty but does not prohibit private appropriation. Hence, private entities may appropriate area in outer space or on a celestial body, although states may not. Under the common law theory of property rights, however, states (lacking sovereignty), would not have any rights to confer on private entities. Conversely, under the civil law view, property rights would exist independent of sovereignty, and therefore could be recognized.[2]

White's interpretation of Article 2 is erroneous because it does not take the language of Article 6 into consideration. Moreover, at the end of White's paragraph, we have the use of the passive voice, which is always useful in obscuring who is doing what to whom: "property rights could be recognized" by whom? Again, there must be some legal regime that provides for property rights, some form and exercise of sovereignty.

In recent years, the difference in the wording of Article 2 of the Outer Space Treaty and other documents of the time, specifically the omission of any reference to private activities, has been advanced as an argument that the treaty was not meant to prohibit private real property rights.

> International lawyers differ in their interpretation of the term "national appropriation." Some interpret Article II narrowly to prohibit only national appropriation. Many others interpret the clause broadly to prohibit all forms of appropriation, including private and international appropriation. When Article II is compared to similar provisions in other documents, however, it becomes clear that the narrow interpretation is correct.
>
> Before the Space Treaty was drafted by the U.N. Committee on the Peaceful Uses of Outer Space (COPUOS), four other international legal organizations prepared draft resolutions. All of these documents recommended non-appropriation clauses which are broader than Article II. The terminology in these clauses suggests that at the time the Space Treaty was drafted, international lawyers did not consider "national appropriation" to be an all-inclusive phrase."[3]

As an example, White cites Article 3 of the March 15, 1964, Draft Resolution of the International Institute of Space Law as having specifically distinguished between national and private appropriation:

> Celestial bodies or regions on them shall not be subject to national or private appropriation, by claim of sovereignty, by means of use or occupation or by any other means.[4]

Because the IISL document is the only one White cites, his case is far from compelling. One can effectively argue that language against private appropriation exists elsewhere in the treaty. It is unnecessary for Article 2 to specifically mention private appropriation because the phrase "by any other means" covers private appropriation. The clear intent of Article 6 is to include appropriation by nongovernmental entities as one of the "other means" of appropriation that is prohibited. When White suggests that "international lawyers did not consider 'national appropriation' to be an all-inclusive phrase" that included private appropriation, we can agree with him and at the same time understand the specific distinction between "national and private appropriation" in the IISL document as being no different from the distinction between "claims of sovereignty" and "other means" in the Outer Space Treaty.

Carl Q. Christol, professor of international law and political science at the University of Southern California, offers a very different interpretation of history regarding the 1964 IISL draft. In Christol's view, although there was support for the IISL language, including the reference to private appropriation, there was a growing awareness of humanity's ability to have access to the space environment and the needs to make use of its resources. From this emerged an unwillingness by states to impose arbitrary legal constraints on either public or private access and use, exploration, and exploitation. Christol recounts the transition from the 1964 IISL language to the language of the 1967 Outer Space Treaty. In 1966, M. Smirnoff of Yugoslavia submitted a proposal that included the prohibition of private appropriation; however, the proposal also included a provision that nongovernmental entities "may explore and use" celestial bodies with the permission of the parent state. As Christol described:

> Thus, the critical distinction was made between the non-establishment of private property rights in the form of appropriation, e.g., exclusive use, and the private right to engage in non-exclusive uses in the process of exploration. But, this distinction was made regarding the celestial bodies, per se, and not with regard to their natural resources.[5]

There is evidence that "national appropriation" was understood as an all-inclusive phrase. Thus, to also mention "private appropriation" would have been redundant:

> The preliminary work on the Treaty clearly shows that several delegations interpreted—apparently without contradiction—as testified by Belgium (A/AC.105.C.2/SR.71 and Add. 1, 4 August 1966)—the term "national appropriation" as covering *both the establishment of sovereignty*

and the creation of titles to property in private law. Civil law property cannot be understood without a public law framework founded on the existence of a State's administrative and coercive power. Private law ownership cannot be thought of without a system of material or territorial jurisdiction of a State. On the other hand, territorial jurisdiction forms the essential contents of national sovereignty. Since—under Art. II of the 1967 Outer Space Treaty—sovereignty is prohibited on other celestial bodies, it goes without saying that outer space and celestial bodies are not only not subject to national appropriation, but also to appropriation under private law.[6]

A broad reading of the intent of Article 2 is bolstered by predecessor documents, including United Nations General Assembly Resolutions 1721 and 1962 (the International Cooperation Resolution and the Declaration of Legal Principles, respectively). Further, these preceding resolutions must be understood in the context of resolutions adopted by other international law fora at the time. For instance, the resolution of the International Law Association adopted at its Hamburg conference on August 13, 1960, recommended "the conclusion of an international agreement whereby states would agree not to make claims to sovereignty or other exclusive rights over celestial bodies." Obviously, real property rights as commonly applied on Earth are exclusive rights, and as such, these rights were to be prohibited on the Moon and other celestial bodies. Resolution 1721, adopted a year later, while not identical in its specific language, was not substantially different. Similarly, the resolution adopted by the Institute of International Law at its Brussels conference on September 11, 1963, declared "celestial bodies are not subject to any kind of appropriation." These slight differences in language from the treaty language are not substantive. These documents resonate with the same tone. Despite this, White concludes:

> Ultimately, Article II must be interpreted narrowly. For under international law states may do whatever is not expressly forbidden. "Restrictions upon the independence of states cannot . . . be presumed." [PCIJ 1927] The language in Article II "by claim of sovereignty, by means of use or occupation, or by any other means," refers to the traditional (occupation) and the broader modern (display of authority) standards for establishing territorial claims.[7]

This does not logically follow. Again, White chooses not to address Article 6, a fatal flaw of omission in his argument. Article 6 expressly states that states parties bear international responsibility for the activities of nongovernmental entities, and these are among the "other means" by which national appropriation is prohibited. White continues to adhere to his position in a later article:

> Article II of the Outer Space Treaty prohibits territorial sovereignty but does not prohibit private appropriation. Hence, private entities may appropriate area in outer space or on a celestial body, although states may not.[8]

Again, this argument conveniently ignores the "any other means" clause of Article 2, as well as its tie to Article 6, which obligates states to ensure that national activities, including those of nongovernmental entities, are carried out in conformity with the provisions of the treaty. Thus, states have a duty to revoke the license of a national entity, or entity launching from its territory, that violates provisions of the treaty. As space law specialist Lawrence A. Cooper states:

> Some have argued that [Outer Space Treaty]'s broad definitions allow individual appropriation of space and celestial bodies because it only specifically prohibits appropriation by States; however, States are responsible for the actions of individuals, and property claims must occur through the State's property laws. Therefore individuals may not claim space or celestial bodies.[9]

Another author on space law, Leslie I. Tennen, concurs:

> The assertion that article II of the Outer Space Treaty does not apply to private entities, since they are not expressly mentioned therein, must fail for the simple reason that private entities do not need to be expressly listed in article II to be fully subject to the non-appropriation principle. . . . The Outer Space Treaty did not create a dichotomy in this regard between governmental and non-governmental activities in space, but rather established the basic principles upon which all space activities, public and private, are to be conducted. Non-governmental entities, as discussed above, must be authorized to conduct activities in space by the appropriate state of nationality. States do not have the authority to license their nationals, or other entities subject to their jurisdiction, to engage in conduct which is prohibited by positive international law to the state itself.
>
> The validity of this principle can be demonstrated by applying the same rationale to other articles of international instruments. That is, if a state may authorize its nationals to "privately appropriate" areas of the Moon and other celestial bodies, notwithstanding article II of the Outer Space Treaty, then it must be posited, why the state could not also authorize its nationals to conduct other activities, in their capacity as private entities, in contravention of other articles of the Treaty or any other international instrument. . . . The illogic of the argument that private entities are not subject to article II, carried to its ultimate conclusion, would negate every bilateral or

multilateral agreement ever made. States could engage in every activity they agreed to restrict or limit by the convenient subterfuge of conducting the activity through the guise of the private, rather than the public, sector.[10]

Economist Sam Dinkin, who advocates the development of real property rights in outer space, likewise believes that they do not exist under current treaty language:

> The Outer Space Treaty of 1967, which has been ratified by 98 nations and signed by an additional 27, forbade property rights in space. No nations can make property rights claims. Further, the conventional interpretation of the treaty is that no one at all can make property rights claims.[11]

In addition to considering arguments based on Articles 2 and 6, it should be noted that Article 1 states that

> [o]uter space, including the moon and other celestial bodies, shall be free for exploration and use by all States without discrimination of any kind, on a basis of equality and in accordance with international law, and there shall be free access to all areas of celestial bodies.

Shin Hongkyun has pointed out that "[a]ppropriation of vast tracts of land for their exclusive use violates Article I, and is unnecessary to ensure non-interference in the vicinity of an activity."[12] Certainly the appropriation of Alaska-sized and even United States–sized territories as advocated in Alan Wasser's Space Settlement Initiative is antithetical to "free access."[13]

Some space property rights advocates, such as Arjen F. van Ballegoyen, would circumvent this inconvenient treaty by unilaterally reinterpreting it:

> Article 2 of the treaty . . . needs to be interpreted in a restrictive, literal meaning, namely as just the prohibition of national appropriation. This interpretation would allow other entities like private companies and non-governmental organizations to appropriate territory.[14]

Of course, this is no legal remedy at all. Van Ballegoyen's idea of arbitrary interpretation makes a mockery of legal principle. The Law of Treaties, Article 31, paragraph 1 states:

> A treaty shall be interpreted in good faith in accordance with the ordinary meaning to be given to the terms of the treaty in their context and in light of its object and purpose.[15]

For a state to unilaterally reinterpret a treaty long after becoming party to it would certainly be unusual, possibly unprecedented, and would have little hope of being internationally accepted. Practically speaking, such an act would be tantamount to abrogating the treaty.

Even White does not intend to assert a theory of unlimited real property rights, as he goes on to clarify:

> Although proponents of space development would undoubtedly welcome the economic incentive of unlimited appropriation, such claims should not be recognized. This form of property rights could potentially preclude free access to outer space in the same manner as territorial sovereignty would preclude free access. Finally, as a point of law, recognition of real property rights beyond the confines of space facilities would be inconsistent with the common law theory of property.[16]

THE NECESSITY OF REAL PROPERTY RIGHTS

Do for-profit ventures need to own real property to extract resources? Van Ballegoyen thinks so:

> [W]e have to come up with a more appropriate regime. Such a regime would have to include the possibility of acquiring ownership of the territory itself. This is the only way to increase the available level of incentives.[17]

Or is it? A few years after the Outer Space Treaty's entry into force, international law professor Marco G. Markoff insisted:

> Appropriation of parts of planets, or of planetary nonrenewable natural resources, is not to be allowed to private entities too. While activities of governmental or nongovernmental bodies were considered, under Art. VI and VII of the Treaty, as national activities, private appropriation of resources has not been prohibited expressis verbis by Art. II. [However,] private appropriation cannot be conceived apart from a public law asset having the power for protecting it. Such an asset is namely the state jurisdiction over a territory. One cannot imagine private appropriation as independent of legal authority of a State. Since no State is entitled to extend administrative or judicial authority over planetary areas beyond the sites where bases or stations are established, no legal basis remains to individuals for occupying, or appropriating by means of private law categories, parts of planets, or their resources.[18]

Markoff's basic premise is sound: where there are no laws, there are no rights. However, he extrapolates too far. Although it is true that "no State is entitled to extend administrative or judicial authority over planetary areas beyond the sites where bases or stations are established," Article 6 requires that

> States Parties to the Treaty shall bear international responsibility for national activities in outer space, including the Moon and other celestial bodies, whether such activities are carried on by governmental agencies or by non-governmental entities.

Logically, in order to fulfill that responsibility, states must extend administrative and judicial authority over planetary areas in the immediate vicinity of the sites where bases or stations are established. Article 8 make this explicit:

> A State Party to the Treaty on whose registry an object launched into outer space is carried shall retain jurisdiction and control over such object, and over any personnel thereof, while in outer space or on a celestial body.

Where there is a jurisdiction, there is the potential for a right. The right of use is found in Article 1:

> Outer space, including the moon and other celestial bodies, shall be free for exploration and use by all States without discrimination of any kind.

Markoff argues that

> [u]se forms part of an exclusive right of a public or private nature. If not qualified by the norm contained in Art. I, para. 1, the use of planetary areas— including natural non-renewable resources—could easily lead to the concept of a complete public (territorial jurisdiction) or civil (property) exclusive right. The former has been expressly prohibited; the latter is logically excluded under this prohibition. Otherwise, it could turn out to be a disguised feature of national appropriation. For this reason, use cannot be entirely "free."[19]

There is a certain distinction that must be made here. Claims over territory, either public or private, are prohibited; however, the use of resources is permitted, although it "cannot be entirely 'free.'" Herein lies the crack that Markoff seems to avoid prying open. Use cannot be entirely free, in that it cannot take the nature of an extensive and perpetual use right, for such a right would be indistinguishable from a permanent claim over territory. Use is defined by the duration of the operation of a station on a celestial body and to its vicinity. Thus the use right exists within a limited area while the station operates and terminates when the operation ceases. The question that remains is what limits should define a station's vicinity, but this question would only need to be asked in a case where another station operated by another entity might possibly interfere. Article 9 of the Outer Space Treaty provides a mechanism for consultations, which could lead to agreements on limits developing as customary law.*

As with Markoff, Silvia Maureen Williams, professor of space law at the Universidad Nacional de Buenos Aires, seems to rest her position on a perceived norm expressed in Articles 1 and 2, rather than on a logical analysis of the ramifications that flow from the specific provisions of the Outer Space Treaty:

> Although many noteworthy jurists have voiced their thoughts to the contrary, we are more inclined to believe that, in the face of the 1967 Treaty, the intention of its drafters has been none other than to fully ban national appropriation.
>
> We therefore agree entirely . . . with the Argentine doctrine, in the sense that, in view of the norm of positive law represented by Articles I and II of the Space Treaty, appropriation of space resources is banned. It has been upheld that the appropriation of the natural resources forms part of the free-dom of exploration and use of outer space, as is the case in the high seas. However, no rule of positive law exists so far banning the appropriation of resources of the high seas nor imposing an obligation to share the yields. The situation differs completely as regards space resources where, as said before, the 1967 Treaty is positive law on the matter and gives rise to such obligation. The principle of "first come, first served" would have only been applicable in the absence of such norm of positive law."[20]

This is the same type of error made by those who claim that the Outer Space Treaty permits private claims of real property. In this case, however, the error is made to support the other extreme, that no appropriation of any kind is permitted, not even of resources. In either case, the error is in focusing on one or two convenient clauses in the treaty and ignoring other, inconvenient clauses, thereby failing to appreciate the tension between competing principles and find the middle way. It can be asserted with equal effect that the Treaty is positive law on the free use of outer space and its resources. So, just as Markoff observes that "use cannot be entirely 'free,'" in interpreting the entirety of the treaty we must conclude that appropriation cannot be entirely banned.

What was the intention of the drafters? Nathan C. Goldman of South Texas College of Law provides historical context for Article 1:

> From the beginning, the U.S. position has been that in the Outer Space Treaty, Article I, "use" includes resources. In negotiations on this language,

---

* Also, presumably, one of the functions of the ill-fated Moon Agreement's envisioned regime would have been to operationalize a dispute settlement system.

the French and the Hungarian representatives likewise acknowledged that "use" included "exploitation" of resources and other attributions and applications of outer space.[21]

The customary law that has developed since 1967 confirms this principle and falsifies the sweeping statements against resource appropriation asserted by Markoff and Williams in 1970. Ownership of lunar samples by the United States and the Soviet Union/Russia has never been seriously challenged. Furthermore, some lunar material (sold by the Russian government) is now privately owned. Two decades after U.S. and Soviet missions returned resources from the Moon, Gennady M. Danilenko, a Soviet expert on international law, asserts:

[T]he Outer Space Treaty proclaims freedom in the use of outer space, which, as generally recognized, includes the freedom to exploit its resources.[22]

White concurs:

[A]lthough entities may not claim ownership of mineral resources "in place," once they have been removed (i.e. mined) then they are subject to ownership.[23]

The U.S. delegation stated in U.N. Committee on the Peaceful Uses of Outer Space (COPUOS), at the time that the Moon Agreement was negotiated, that the words "in place" allow private property rights to apply to extracted resources.[24] That statement went unchallenged. Although the Moon Agreement is not binding on the United States (nor on most other states), the delegation's unchallenged statement is additional evidence of customary law on this point.

Another *res nullius* was negotiated concurrently with the Moon Agreement. The United Nations Convention on the Law of the Sea (UNCLOS III), Annex III, Article 2 states:

Title to minerals shall pass upon recovery to the entity which mined them.[25]

Because the UNCLOS III has been widely accepted by the international community and much of space law derives from maritime analogies, this principle can be applied to outer space as a point of customary law.

Finally, the Russian Federation has set the precedent of incorporating this principle into its municipal law:

The property rights over the physical product created in outer space shall belong to the organizations and citizens possessing property rights in the components of space technics.[26]

As a matter of legal theory, a property right to a resource is created when labor is mixed with soil. Thus, the real issue is the right to extract or "use" the resource, which Article 1 of the Outer Space Treaty permits, and over which states shall exercise jurisdiction pursuant to Article 8. This, together with the right to be free of interference, contained in Article 9, creates rights that exist while use and occupation are ongoing. Although the transitory and limited nature of state jurisdiction means that no permanent real property right can be granted, claimed, or recognized, this seems unnecessary.

The Outer Space Treaty does not forbid in situ resource utilization. Space is treated like a commons. Astronauts have brought home space rocks and taken title to them. If you want resources on Mars or the Moon, take them.[27]

To this I would add the argument that if it were profitable for a company to go to the Moon and pick up rocks, it would. It is not profitable at this time. So, how is it profitable for the company to claim title to the land for miles around the rocks that are too unprofitable for it to pick up, land that contains yet more unprofitable rocks? Outer space is to be "free for exploration and use by all States," and by devolution, by the private parties of all states. What would be the point of private property rights in the traditional sense? If a private party has the means to go to the Moon or other celestial body and to use the resources there, it is free to do so. No property claim is necessary. Only if more than one private party were interested in exploiting the same site on the Moon or Mars or an asteroid would there be a conflict that would need to be adjudicated by an international space authority. Even here, the rights to be adjudicated would be limited in nature, pertaining to specific uses and duration, rather than rights to real property in perpetuity.

Finally, Earth is replete with examples of private, for-profit activities on public land, such as livestock grazing and timber harvesting. There are also examples of such private activities in the *res communis* beyond national boundaries, such as the extraction of petroleum in maritime Exclusive Economic Zones. ExxonMobil and BP Amoco do not own the continental shelves, nor do they need to. Private activity on U.S. public lands is based on permits. Today these are granted with few conditions and constraints, but legal doctrines regarding use of public lands in most the legal systems include government licensing of access and regulation of activity. When ExxonMobil, BP Amoco, or others exploit continental shelf resources, they get permits from the coastal state that convey use rights, extraction rights, and protection against "claim jumpers." Such protection is something private firms strongly desire and typically seek from governments rather than try to supply for themselves.

The promulgation of unlimited and perpetual real property rights is a misguided attempt at "the transplantation of archaic political and legal features from the earth on the planets."[28] The consistent position of the United States regarding ownership of extracted resources, as well as the customary law evolving from the ownership and sale of retrieved lunar resources, and the fact that private parties are not subjects of international law in general nor of the Outer Space Treaty in particular, fully protect any private interests that may seek to extract resources from celestial bodies for profit. Although as a hypothetical case, Argentina might vigorously object to a U.S. company mining the Moon, a complaint would have no standing either in U.S. municipal law or in international law; it would be unable to bring suit against the private party except under whatever municipal law it might pass, and then only if the company had a subsidiary incorporated in Argentina that was materially involved in the lunar venture. Presumably, any company smart enough to extract lunar resources at a profit would also be smart enough not to base its operations in a country that insists on a strict interpretation of the "common heritage of mankind" principle. Essentially, the same principle of international law that prevents the United States from granting or recognizing real property claims in outer space also shields U.S. citizens from international disputes over extraterrestrial resource extraction: it is the state party, not the private interest, that bears international responsibility under the treaty.

FUNCTIONAL PROPERTY RIGHTS

University of Mississippi law school chairman Stephen Gorove wrestles with two competing principles in the draft Moon Agreement, which was still under negotiation at the time he writes:

> What makes the prohibition [of property claims] in the new draft somewhat illusory, if not illogical, is the fact that it also stipulates that the placement of space vehicles, equipment, facilities, stations and installations on or below the surface of the moon, including structures connected with its surface or sub-surface are not to create a right of ownership of parts of the surface or sub-surface of the moon or other celestial bodies. This stipulation in fact means that states or organizations could establish facilities, stations and installations on the moon or other celestial bodies and occupy an area over a long period of time or, if human settlement becomes feasible, perhaps even indefinitely, exercising dominion and control over the area subject only to the limited right of visitation guaranteed under the Outer Space Treaty.
>
> Thus it is difficult to see in what manner the draft treaty's prohibition would become effective. States and other organizations, as well as individuals,

could occupy and control the surface or sub-surface of the moon with their vehicles, equipment, facilities and installations, as long as they wished. They could exercise control over it subject only to the aforementioned right of restricted visit. The only thing, therefore, that the draft does is to say that such possession and control will not create a right of ownership over parts of the surface and sub-surface of the moon. But it seems that everything could be exercised by the state, organization or individual much the same way as if such a right of ownership did in fact exist.[29]

Not quite. Such dominion and control could be exercised by the state, organization, or individual only over the area of occupation or use, not over extensive areas claimed but neither occupied nor used, and only for the period of occupation or use, not in perpetuity. With this correction, what Gorove has described are functional property rights that are localized and limited to the duration of occupation and use, as distinct from real property rights that are extensive and perpetual.

Gorove claims that the Moon Agreement contains a prohibition of "grant, exchange, transfer, sale or purchase, lease, hire, gift, or any other arrangement or transaction with or without compensation relating to parts of the surface or subsurface of the moon or other celestial bodies," and Damodar Wadegaonkar concurs.[30] This follows from a reading of a sentence in Article 11, paragraph 3:

> The placement of personnel, space vehicles, equipment, facilities, stations and installations on or below the surface of the moon, including structures connected with its surface or subsurface, shall not create a right of ownership over the surface or the subsurface of the moon or any areas thereof.

What does not follow from this provision is that "space vehicles, equipment, facilities, stations and installations on or below the surface of the moon, including structures connected with its surface or subsurface" cannot themselves be objects of transaction. Article 12, paragraph 1 provides:

> The ownership of space vehicles, equipment, facilities, stations and installations shall not be affected by their presence on the moon.

Article 8, paragraph 3 provides:

> Activities of States Parties in accordance with paragraphs 1 and 2 of this article shall not interfere with the activities of other States Parties on the moon.

The right to noninterference requires an effective zone of control if it is to have any meaning. Because "space vehicles, equipment, facilities, stations and installations" can be owned, they can be transferred in the same

manner as any other form of property. When a new entity assumes ownership of such assets, it also acquires control of the zone necessary to
assure noninterference with the operation of those assets.

White makes the point that although the Outer Space Treaty states that
space "is not subject to national appropriation by claim of sovereignty,"
it also obligates states to exercise other forms of state sovereignty, particularly jurisdiction over objects that they or entities under their jurisdiction launch into or construct in space. This jurisdiction permits the state
of registry to subject its space objects and personnel to any national laws
that are not in conflict with international law. States may legislate with
respect to a broad range of both public and private activities; and, in
most circumstances, they exercise as much authority within the vicinity
of their space facilities as they would within their territory on Earth.

> Under Article VIII, jurisdiction and control is only valid insofar as it is
> necessary to accomplish the exploration and exploitation of outer space and
> celestial bodies. Jurisdiction and control is also limited in time. It ceases to
> exist when activity is halted—as, for example, when a space object is aban-
> doned or returned to Earth. Because states only control as much territory as
> is actually used, the Outer Space Treaty does permit free access to outer
> space.[31]

White describes this as limited "functional sovereignty," which is
distinct from unrestricted "territorial sovereignty." This concept provides
the basis for a limited and temporary form of property right. While one
is on the Moon or Mars or an asteroid, one has the right to use the immediate area free from harmful interference, and one has the right to use
resources in situ or to transport them elsewhere, including bringing them
to Earth. He notes that Csabafi proposed that "'designated zones' of
functional jurisdiction would permit unilateral action in outer space."[32]

> Csabafi suggests the necessity of an international agreement that "would
> define certain specific cases when a state, being able to show a 'particular
> and distinctive interest,' may claim the right to exercise functional jurisdic-
> tion in a designated zone of outer space or on a celestial body." States would
> then create "designated areas" of functional sovereignty through unilateral
> legislation. Csabafi analogizes to the regime on the continental shelf, and to
> the functional sovereignty which some nations exercise over pearl and
> sedentary fisheries on the seabed. Unfortunately, the zones which Csabafi
> describes are ill-suited to the complex interactions which will occur when
> industry and habitation become routinized.[33]

White does not explain this last statement, nor does he explain why
"[i]n general, real property law would seem to provide more appropriate

analogies when addressing the problems associated with permanently located space facilities and mining sites." This might be true in the remote future, when Mars is crowded with millions of people, but it hardly seems relevant to the near-term situation. Mars has the same dry land area as Earth, but with the important difference of having seven billion fewer people. Surely there is no immediate need to draw property lines in the Martian or lunar dust.

In any case, pursuant to the "free access" language in Article 1 of the Outer Space Treaty, White concludes that property rights could only be valid within the immediate confines of a facility. From this, it follows that Dennis Hope's Lunar Embassy scheme is without legal foundation, as is Alan Wasser's Space Settlement Prize Act. One does not have title to what one does not occupy and use; therefore one cannot convey title to others.

With regard to the space within a facility, there is little question that international law permits property rights. However, as White points out:

> [I]n light of the maxim that entities cannot transfer a greater right than they have, these property rights would be, in common law jurisdictions, necessarily more limited than traditional property rights. The common law sovereign could only confer title to the extent of its own sovereignty; thus, under the functional sovereignty conferred by Article VIII of the Outer Space Treaty, property rights would be functionally defined and limited in time.[34]

White then applies the concept of functionally defined and temporally limited property rights to the immediate vicinity of a facility:

> Under a regime of functional property rights, title would arise on the basis of a principle entirely different from traditional property rights. Conferral of title would not depend upon a government's control over a specific area, but rather upon its control over the space objects and personnel at that location. Once conferred, these rights would, nevertheless, be almost identical to terrestrial property rights.[35]

Thirty years earlier, U.S. Attorney General Nicholas Katzenbach referred to "primary rights . . . in a localized facility" that exist by virtue of the activity ongoing in the facility, independent of any consideration of real property ownership.[36] White applies this idea to outer space activities:

> [S]tates may legislate with respect to a broad range of both public and private activities; and, in most circumstances, they exercise as much authority within the vicinity of their space facilities as they would within their territory on Earth.

In space, first-come, first-served occupation, and the prohibition against harmful interference with other states' activities provides states with a similar, albeit less clearly defined, right of exclusion. . . . Functional property rights would be subject to the limitations of [Outer Space Treaty] Article VIII jurisdiction. These rights would terminate if activity were halted, as for example, if a space object was abandoned or returned to Earth. Finally, rights would be limited to the area occupied by the space object, and to a reasonable safety area around the facility.[37]

Article 8 provides:

A State Party to the Treaty on whose registry an object launched into outer space is carried shall retain jurisdiction and control over such object, and over any personnel thereof, while in outer space or on a celestial body. Ownership of objects launched into outer space, including objects landed or constructed on a celestial body, and of their component parts, is not affected by their presence in outer space or on a celestial body or by their return to the Earth.

But how does this jurisdiction translate into "functional property rights . . . around the facility?" Article 9 states that

[i]f a State Party to the Treaty has reason to believe that an activity or experiment . . . would cause potentially harmful interference with activities of other States . . . it shall undertake appropriate international consultations before proceeding with any such activity or experiment. A State Party to the Treaty which has reason to believe that an activity or experiment planned by another State Party . . . would cause potentially harmful interference . . . may request consultation concerning the activity or experiment.

White's theory of functional property rights in outer space is intriguing, although his case is not as strong as it could be because he ignores the provisions of Article 6. It is necessary to state that Article 6, which applies Article 2 to all entities under state jurisdiction, does not prohibit functional private property rights because the limited duration and extent of these rights does not constitute "national appropriation . . . by means of use or occupation, or by any other means"; rather, the stronger tie is to the resource use right in Article 1, to the state's jurisdictional obligation over operations (including resource use) in Article 8, and an operation's implied right to a limited zone of safety and noninterference in Article 9. White explains that

because the safety zone jurisdiction outside facilities would be strictly limited, entities would not be claiming large areas. This means that different facilities and their safety zones could each be under the jurisdiction of a different state, and yet still be in close proximity to each other.[38]

To the extent that such rights of limited duration and extent are compatible with the Outer Space Treaty,

> [t]he regime is attractive because it is so easy to implement. Nations can unilaterally enact legislation, and they can tailor that legislation to conform to their existing property laws.[39]

In the absence of any new international agreement, with the language of the Outer Space Treaty as a springboard, functional property rights might develop as customary law as the national entities of launching states extract such extraterrestrial resources. However, in his next paragraph, White describes the international regime that states should establish for defining in greater detail and standardizing functional property rights:

> Participating states should additionally provide for reciprocity and/or negotiate some form of limited "mini-treaty" to coordinate national property legislation. Such a treaty would elaborate on the elements in Article VIII—it would define the property rights conferred under Article VIII ... it would delineate the extent of jurisdiction and control, with particular emphasis on the physical extent of safety zones, and upon the temporal duration of jurisdiction, i.e. upon the period of abandonment necessary to extinguish jurisdiction.[40]

THE CONSTITUTION OF OUTER SPACE

The official statements of members of the Bush administration and quotes from space enthusiasts provides an instructive study in contrasts. From Norman P. Neureiter, science and technology adviser to the secretary of state:

> [T]he Outer Space Treaty and three related UN conventions ... serve as the bedrock of international space law. This was an example of multilateral diplomacy at its best; the international rules that were created afford a measure of transparency and accountability for space activities, without constraining national programs.[41]

From Ambassador Kenneth Brill, permanent representative of the United States of America to the United Nations in Vienna:

> This 35th anniversary of the Outer Space Treaty is also an opportunity for us to address the fact that the world is far from general acceptance of the four core space law instruments: the Outer Space Treaty, the Rescue and

Return Agreement, and the Liability and Registration Conventions. Several key States have not accepted key treaties, including some members of COPUOS. This Subcommittee should make a clear call for States to ratify and implement the four core space law instruments cited above. And, of course, it should encourage States that have accepted the core instruments to look at the sufficiency of their nation's laws to implement them. Parties ought to ensure that they are indeed doing what they have promised they will do.[42]

From Kenneth Hodgkins, U.S. adviser to the fifty-seventh session of the UN General Assembly:

The Outer Space Treaty was in many ways the foundation of the now well-established field of space law and it set the framework and cooperative tone for tremendous technological progress in outer space activities. In no small part, these accomplishments can be attributed to the role of COPUOS and its Legal Subcommittee. Under this legal regime, space exploration by nations, international organizations and, now, private entities has flourished. As a result, space technology and services contribute immeasurably to economic growth and improvements in the quality of life around the world. The Outer Space Treaty has truly stood the test of time; its provisions remain as relevant and important today as they did at the inception of space exploration.[43]

These statements by representatives of the most unilateralist U.S. administration since World War II express the depth of the U.S commitment to the Outer Space Treaty. If the Bush administration backed the treaty to the hilt, certainly a more multilateralist Democratic administration would do no less. Thus there is virtually no chance that the United States will ever undermine the treaty by passing national legislation contrary to its principles, as Wasser advocates with his Space Settlement Initiative. Meanwhile, the disregard for international law expressed by some space development advocates is truly shocking. From Alan Wasser, chairman of the Space Settlement Institute:

Some suggest the U.S. should [opt out of the Outer Space Treaty] . . . I would personally like to see that happen.[44]

The online publication Reason reported:

Jim Benson plans to declare ownership of an asteroid orbiting between Earth and Mars. And he doesn't much care what the United Nations has to say about it. "If the U.N. doesn't like it, they can send a tank up to my asteroid, which of course they can't."[45]

The late Jim Benson was founder and former CEO of SpaceDev. He also appeared to be entirely ignorant regarding the position of the United States and the United Nations with respect to private property ownership in space, and of their legal jurisdiction:

> I don't believe they have an official position, and if they did, I wouldn't care because I don't believe they have legal standing in space—they are earth-based.[46]

In fact, sending a tank up to "his" asteroid would not have been necessary. The U.S. Department of Justice could have brought suit against SpaceDev in federal district court right here on Earth.

Dinkin has called repeatedly for withdrawal from the Outer Space Treaty:

> If bilateral agreements and the Outer Space Treaty do not provide an adequate regulatory environment for commercialization and colonization, then perhaps the treaty should be amended or the U.S. should withdraw.[47]
>
> Let's withdraw from the Outer Space Treaty and establish a private property rights regime that opens up a new land rush into space.[48]
>
> The United States should commence international negotiations to amend the 1967 Treaty of Outer Space or withdraw from it.[49]

Finally, there is Representative Tom Feeney's "thought exercise" in futility. Feeney, the Republican representative for Florida's 24th district, which includes Kennedy Space Center, had this to say:

> As a thought exercise, assume that the United States withdrew from the 1967 Outer Space Treaty—as that treaty allows—and stated it would establish and enforce a private property scheme for space-related economic activities. What types of economic ventures would assemble to take advantage of this opportunity? How would their demand for space vehicles, launch facilities, and related technological innovations transform the aerospace industry and Florida's Space Coast? Finally, how would these activities complement NASA's exploration ventures?[50]

The answer to Representative Feeney's questions is that withdrawal of the United States from the Outer Space Treaty would have a chilling effect—like nuclear winter—on private enterprise in outer space, because it would shatter the bedrock of international space law. The entire edifice would collapse. Although the congressman suggests that the United States could "establish and enforce a private property scheme for space-related economic activities," no other space launching state would recognize such a "scheme," and few reputable private enterprises would wish to take the

risk of doing business in such a lawless environment. Representative
Feeney's district would be known as the Space Coast primarily for its
"Space Available for Lease" signs. It should be noted that Feeney is careful
not to actually advocate withdrawal from the treaty, thus his "thought exercise" is mere lip service to laissez-faire. That Representative Feeney has not
even taken the less drastic course than outright withdrawal by introducing
Wasser's proposed legislation is further evidence that his statement is an
exposition of empty rhetoric. If laissez-faire Republican Feeney, a member
of the 110th Congress' Subcommittee on Space and Aeronautics and representative of the Space Coast, will not introduce Wasser's bill, who will?

Danilenko observes, "Expanding space economic activities requires the
creation of a favorable legal framework." It should be obvious that withdrawing from the Outer Space Treaty, which Danilenko calls "the basis for
all subsequent treaties and other legal instruments relating to space activities," is hardly the way to go about this.[51] The State Department is aware
of the possible constraining effect of the current language in the Outer
Space Treaty, and is willing to seek multilateral diplomatic solutions to
update the treaty. This is the right way to get the job done, as Neureiter
comments:

> But this body of law was developed during an era when nearly all space
> activities were carried out by governments. Perhaps it is time to begin
> thinking about whether it will be adequate for the coming era of space
> commercialization.[52]

A final observation regarding the nature of the Outer Space Treaty
should drive home the enormity of advocating withdrawal from it. Such
a withdrawal would not be equivalent to withdrawing from the 1972
Anti-Ballistic Missile Treaty, which was between two states. As the
codification of near-universal principles, the treaty is regarded as the
"constitution" of outer space.[53]

> As discussed in chapter two, *jus cogens* are peremptory norms of general
> law. Although they bear a resemblance to natural law, these norms are ema-
> nations of positive law, reflecting the evolving consensus of the civilized
> world . . . has suggested that the modified *res communis* (space for the bene-
> fit of all humanity) and other principles in the 1967 Outer Space Treaty are
> candidates for the *jus cogens* status.[54]

Jus cogens is Latin for "compelling law." According to Article 53 of the
Vienna Convention on the Law of Treaties:

> a peremptory norm of general international law is a norm accepted and
> recognized by the international community of States as a whole as a

norm from which no derogation is permitted and which can be modified only by a subsequent norm of general international law having the same character.[55]

Legal scholars Sir Robert Jennings and Sir Arthur Watts add:

Presumably no act done contrary to such a rule can be legitimated by means of consent, acquiescence or recognition.[56]

Jus cogens principles cannot be circumvented by withdrawing from the treaties that codify them. Thus, if the Outer Space Treaty is *jus cogens*, the United States would be bound by its principles even if it withdrew from the treaty.

Some U.S. international lawyers dislike *jus cogens* arguments. In any case, the real possibility remains that if the United States withdrew from the Outer Space Treaty, the other parties of the treaty would act to defend it and could make life difficult for U.S. nationals who conducted activities that those parties regarded as inconsistent with the treaty. It should not be necessary to remind anyone that the United States is not the only launching state.

The question of withdrawing from the treaty should also be considered in the light of a legal doctrine of an objective regime. It is possible for rules and organizations created under a treaty to have a reality and legitimacy with regard to nonparties, because the parties are making them work and creating legally valid results as they act pursuant to those rules or through those organizations.

FANFARE FOR THE COMMON LAW

United Societies in Space, Inc. (USIS), is a Colorado nonprofit corporation formed on August 4, 1994, at a conference in Cuchara, Colorado, by Russians and Americans. It is led by Oleg Alifanov, Ph.D., of the Moscow Aviation Institute, and Declan J. O'Donnell, a former president of the World Space Bar Association, and publisher of the *Space Governance Journal*. According to O'Donnell,

[b]riefly, the mission of USIS and affiliates is to help create a space faring society during our lifetime. We focus on the space policy problems, legal matters, and societal concerns. Our style is to supplement the U.N. International Treaty Regime with functional entities where appropriate people can assemble, work, and actually execute an effort for the benefit of Humankind and its settlers in outer space.[57]

Although O'Donnell finds an "anti-development philosophy" in international space treaty law that I do not, I welcome his initiative as the epitome of the overused phrase, "thinking outside the box."[58] Not only has he thought in the black space outside of Earth's atmosphere, he has thought in the "white space" within the international treaty regime, the legal and administrative void that it has left unaddressed:

> There is a de facto void in statutes, regulations, and rules applicable generally in the territory known as outer space. There is also a de jure void in legal authority generally applicable in outer space in that the five space treaties enact a space policy that prevents Nations from extending their own sovereignty into space. Space treaty law is part of International Treaty Law, but literally legislates the maintenance of a political void in outer space.
>
> The space faring Nations are not able to correct this space policy because they have all signed and ratified a treaty that prevents them from asserting their sovereignty into space.[59]

Note that the last treaty to be widely ratified was the 1975 Registration Convention. Thus, the international community has not only committed itself to maintaining the political void in outer space, it has also neglected the legal and administrative voids.

> [A] legal void in space government was codified. The only exception by treaty law is that Nations may assert their own sovereignty inside of their space ships. The treaty law in this regard is very specific about the applicability of mission rules inside of the space ship so there is no basis for extending those rules outside in order to create property rights on the Moon, Mars, or in orbit.
>
> Therefore, there are no private property estates in the territory of outer space. Instead, space resources are considered common property to be held for the benefit of future generations of humankind, not able to be appropriated by any nation, and not subject to ownership by any person, company or association.[60]

To fill these voids, O'Donnell proposes a new paradigm in space governance:

> The Regency of United Societies in Space (ROUSIS) ... is a common law government trust with a character of compliance to space treaty principles. The Regents are trustees and Humankind is the beneficiary. Space resources are the *res* of this trust. The constitution has extended common law into outer space not only for the purpose of accommodating this new citizen movement entity, but, also, for the purpose of establishing a basis for law and order among settlers.

There is no intent nor legal capacity to replace treaty law. Instead, the idea is to supplement it with workable and equitable solutions to every day space settlement problems.[61]

The key concepts here are "compliance to space treaty principles" and "to supplement ... treaty law." Thus, the ROUSIS seeks to exist in harmony with existing international law. O'Donnell draws his inspiration from the tradition of English common law and courts of equity, which the United States inherited:

> Common law remedies were available only while the Kings and Queens of England, or the Congress of the United States here in America, did not object and there was no remedy at law under their dominant paradigm. Only then could the Common Law Courts of Equity grant an equitable remedy to supplement the legal system in place.[62]

O'Donnell's ROUSIS recognizes the sovereign states of the international system as "the kings and queens," and the treaty regime as the sovereign law. So long as the sovereigns do not object and there is no remedy at international law, ROUSIS arrogates the authority to grant an equitable remedy to supplement the legal system in the character of a common law court of equity.

> The first principle of common law is that no remedy exists "at law" i.e. in the King's Court. If one is available, or if the King intervenes in any way, the common law yields to that authority. The United Nations and its member nations are analogized to the King and the King's Courts.
>
> The second principle of the common law property estates is that they relate only to temporary usage on a fair and "equitable basis." They exclude any concept of legal ownership. However, in modern times, the equitable estates represent the core of most commercial transactions. The legal title known as "fee simple absolute" is rarely relied upon by businesses, although it is available at law.
>
> A non-sovereign Regency of United Societies In Space, (ROUSIS), is planned for adoption on August 4, 2000. A minimum of 200 Regents would organize this 100 year entity to provide legislative, executive, and judicial departments for space governance purposes. It would be charged with the duty to transform humanity into a space faring society, build an appropriate space based infrastructure, and cause a more permanent government with United Nations approval to be established by settlers in space on or before August 4, 2100. During the 100 year Regency authorities for the Moon, Mars and larger orbits would be able to coordinate space resource uses with the regency legislature; resolve disputes in a unique space oriented court system, and obtain executive assistance as requested. If

recognized internationally, this governance structure would also issue space money to pay for space development.

The Regency proposal is designed to cure the most pressing space policy problem of all: It will assist in the governance, jurisdiction, and consensus management of space resources. This will involve the maintenance of a system of private property in outer space, one that is traditional, intelligible, relevant, and possible. It is planned that such system be created well before settlers arrive in space so investors can adjust to it by advancing funds up front, rather than waiting to see if space in fact develops without private enterprise.[63]

Many countries around the world have far more statist legal traditions than the United States. How these countries would react to ROUSIS is an interesting question. In many countries, private activity needs explicit state approval, either as a category of undertaking allowed in law or through individual licensing of particular enterprises.

O'Donnell makes it clear that common law property estates relate only to usage; they do not include any legal ownership. This is a particularly important point, and the great advantage of his concept over others. The nonappropriation treaty language means that common property cannot be owned by anyone. In pointing to existing examples of the common law of outer space, he describes principles that developed as customary law. What international law scholars refer to as "customary law" might be called "international common law"—it is law that exists in the absence of treaty law or sovereign objection:

> In the beginning of space law two principles were established by common law processes. The first is the space law that satellites may trespass over national borders without permission. *Sputnik* took one revolution in orbit in 1957 and no nations filed any formal objection. That established space common law so strongly that it persists today despite the enactment of five space treaties after that, none of which mentioned this law: there was no need to do so because there was no objection.
>
> Another principle of space common law is that of non-interference which is now codified at Article 9 of the Outer Space Treaty.[64]

O'Donnell summarizes the key features of astro law estates under an A-E-I-O-U formula:

A. The rule of property law will apply in the ASTRO venue only.
E. EQUITABLE estates at common law will be used and limited to usage without any ownership.
I. All such user estates, leases, easements, and mortgages, will be INCHOATE and defeasible by treaty law later enacted.

O. All such rules are applicable OUTSIDE of space ships because the Outer Space Treaty Article VIII reserves the inside for the exclusive jurisdiction of the sponsoring states.
U. All of the estates are also subject to USES required by treaty and other UNIVERSAL legal burdens already in existence.[65]

According to the ROUSIS Constitution, Article 2, Section 12:

ROUSIS shall have the following governance powers in order to prepare settlements and municipalities in space for future self-governance.
A. To establish and maintain a body of Common Law estates in property, such as tenancies of 99 years or less, easements, and trust estates, as well as a Common Law of contracts, torts, and criminal laws, as extended to outer space herewith.
B. To establish and maintain a Treasury Department and a monetary and fiscal policy for space; to coin money and regulate the values and uses thereof; to issue money for space governance purposes; to issue notes, bonds, and financial guarantees in return for products, services, and infrastructure which may be inherited by successor government(s) by the year 2100.
C. To establish and maintain standards for weights and measures and language in space, as well as a common standard for keeping time.
G. To establish and maintain a state-of-the-art intellectual property protection system, including, but not limited to, an enforceable patent system for valuable inventions, and a copyright law for unique writings and works of art and music, as well as trademarks and service marks.
H. To license corporations.
P. To establish and maintain a Land, Mineral, and Mining Rights Claims Office.
Q. To establish and maintain a Department of Mining and Energy to facilitate, coordinate, and regulate the research, development, licensing, and operations of resource material harvesting and processing.[66]

Many of these functions are similar to those that might have been provided by the regime envisioned in the Moon Agreement (see chapter 5, "Establishment of a Governing Regime").

Article 2, Section 12, paragraph B of the ROUSIS Constitution even provides for an extraction of value by the ROUSIS government in the form of money, "notes, bonds, and financial guarantees in return for products, services, and infrastructure." In other words, means for the government to finance itself and receive revenue, that is, taxation (see chapter 5, "Death and Taxes").

Article 2, Section 12, paragraph B of the ROUSIS Constitution mentions the establishment and maintenance of a "common standard for keeping time." A new civil timekeeping system will be needed on Mars, where the solar day is 39.6 minutes longer than on Earth; furthermore, given the seasonal changes in the Martian environment, a new civil calendar will also be needed. Of course, these new standards for keeping time will need to be calibrated to Universal Time, Coordinated (UTC) and the Gregorian calendar.[67]

Article 2, Section 1 of Exhibit C to the ROUSIS Constitution provides for legislative standing committees with oversight over corresponding public or semiprivate economic development authorities (EDAs) for development on the Moon, Mars, the asteroids, and in orbit. An EDA would be:

> a likely manager of astro law estates that are based on common law extended into space. It is a trust estate itself and does not contravene any of the Space Treaty provisions on sovereignty, appropriation, benefit sharing, or international cooperation. . . . It would be able to register leases, easements, and mortgages for developers.
>
> . . . we may require that our space industry governance entities, be designated the new regime under the Moon Treaty. This would permit space industry influence in the management of space resources, one that could clarify reasonable and "equitable sharing" rather than "Common Heritage of Mankind" standards. It may lead to the treaty designation of the Lunar Economic Development Authority, Inc., as the new regime for the Moon.
>
> . . . the common law estates should be used in outer space where the United Nations acting as the King's law, has left a void and is silent on the subject. The space governance entities, particularly the Lunar Economic Development Authority at the venue of the Moon, should be able to forge these estates for the benefit of humanity, and all nations, and the United Nations.
>
> In this context the space governance entities are analogous to the trust estate. They feature a court system that may succeed to the tradition of the court of equity. Users of the Moon may register their negotiated common law leases with the relevant space governance entity, such as the Lunar Economic Development Authority (LEDA). Because there is a treaty relating to registration, that entity would cross-reference all of its leaseholds, easements, and mortgages with the proper United Nations office in charge of registration. This office only registers space objects so the LEDA would clarify that its leaseholds are man made personal property intended for usage in space, thus expanding the space law of registration, but doing so in a common law way.[68]

One of the issues that O'Donnell believes the common law of outer space should address is the case of "mixed property:"

> There is a peculiar circumstance that arises in space law when space objects are mixed with space resources. The space treaties have no mention of how to treat the resulting property. For example, if a Lunar orbiting space ship was enlarged by adding cement with Lunar regolith to build a larger vessel, would the resulting hybrid vessel be a space object (subject to forever liability risks to the national sponsor) or would it be a space resource, (subject to forever benefit sharing). No one knows for sure.[69]

I doubt that any part of the resulting hybrid object could be considered a space resource (and if it were, it is highly questionable whether it would be "subject to forever benefit sharing"). It cannot be denied that the portions of hybrid object that came from Earth were once resources, but once the value of human labor is added, and even a highly mechanized or fully automated manufacturing process represents human labor once removed (*Quis fabricat ipsos fabricatores?**),these things become property. According to Article 11, paragraph 3, of the Moon Agreement, only "natural resources in place" cannot be claimed as property. Although the lack of wide ratification renders the agreement as something less than settled law, it is at least evidence of customary law on this point (see chapter 5, "Resource Property Rights"). As for "benefit sharing," the question is not whether it is "forever," but whether it will be *ever.* The concept of sharing benefits appears in many UN resolutions pertaining to outer space, but these are merely declaratory and do not constitute "hard law." Even the Outer Space Treaty is a "Treaty on Principles." O'Donnell emphasizes "reasonable and equitable sharing" over the "common heritage of mankind" principle found in the Moon Agreement. However, the term "equitable sharing" is to be found in Article 11, paragraph 7(4), of the agreement, and since the agreement also states that the "common heritage of mankind" principle finds its expression in the provisions of the agreement, "equitable sharing" is therefore part of that expression. Furthermore, three decades later, the Declaration on International Cooperation states that "States are free to determine all aspects of their participation in international cooperation in the exploration and use of outer space on an equitable and mutually acceptable basis." It is up to states to decide the extent of sharing from their activities. Nevertheless, a clarification in common law would certainly be welcome.

---

* Who makes the makers themselves?

Various media were given five years advance notice of the Denver Space Constitutional Convention. In addition, special notice was given to the aerospace industry. All of the member delegations of the United Nations were given noticed in early 2000.

> Delegates to the Convention assembled at the University of Denver College of Law, International Legal Studies Program, co-sponsored with United Societies in Space. ... The sessions on August 4 and 5, 2000 A.D. were adjourned for one year and until August 4, 2001. This allowed time for conference amendments to the draft constitution to be circulated and ratified one year hence. The ratification process would be completed by vote of 2 to 3 of the Regency's legislative body, as would all future amendments be processed.
>
> The base document was published in *Space Governance Journal*, volume 6, page 22.[70]

NASA appointed an observer, Mrs. Diana Hoyt, chief of the Department of Policy and Plans, signifying the acquiescence of the U.S. government for the time being. Although newly developed, ROUSIS is a fact. Serving as its vice president is none other than Edwin Aldrin, *Apollo 11* Lunar Module Pilot and the second human to walk on the Moon. ROUSIS not only has the capacity to create common law, but over time, the possibility of serving as a source of customary law for the international legal system.

In this context, the establishment of property rights (but not title in perpetuity) is an expression of popular sovereignty rather than an exercise of national sovereignty. It is a quiet concept that makes no extravagant claims and does no violence to existing legal principles of outer space law.

Though ROUSIS is a fact, at this time it is a tiny fact. Counters on the USIS Web site in February 2006 indicated roughly a hundred visits, give or take a few dozen. Perhaps it is so clever an idea that the space enthusiast community has been slow to understand and embrace it. Perhaps O'Donnell is less inclined to promote his idea than are those who hawk far less worthy concepts. It remains to be seen whether ROUSIS will gain wide enough acceptance to become an effective common law regime for outer space. Also, although former possessions of the British Empire may have inherited the common law tradition, most of the world's nations have not. In addition, in many countries, private activity requires explicit state approval, either as a category of undertaking allowed in law or through individual licensing of particular enterprises. All that can be said at this point is that ROUSIS is a positive and promising project that has been undertaken by credible people.

NOTES

1. White, Wayne N. 1998. "Real Property Rights in Outer Space." *Proceedings, 40th Colloquium on the Law of Outer Space.* American Institute of Aeronautics and Astronautics, 370. Available from http://www.spacefuture.com/archive/real_property_rights_in_outer_space.shtml; accessed March 19, 2005.

2. Ibid.

3. Ibid.

4. International Institute of Space Law. 1965. "Draft Resolution of the International Institute of Space Law Concerning the Legal Status of Celestial Bodies," in *Proceedings, 40th Colloquium on the Law of Outer Space.* American Institute of Aeronautics and Astronautics, 351.

5. Christol, Carl Q. 1980. "The Common Heritage of Mankind Provision in the 1979 Agreement Governing the Activities of States on the Moon and Other Celestial Bodies." *The International Lawyer* 14:429, 448.

6. Markoff, Marco G. 1970. "A Further Answer Regarding the Non-Appropriation Principle." *Proceedings, 13th Colloquium on the Law of Outer Space.* American Institute of Aeronautics and Astronautics, 84–86.

7. White, "Real Property Rights in Outer Space."

8. White, Wayne N. 2004. "Interpreting Article II of the Outer Space Treaty." IAC-03-IISL.2.12. Available from http://www.spacelawstation.com/whiteArtII.pdf; accessed March 18, 2005.

9. Cooper, Lawrence A. 2003. "Encouraging Space Exploration Through a New Application of Space Property Rights." *Space Policy,* 19:111–118.

10. Tennen, Leslie I. 2003. "Commentary on Emerging System of Property Rights in Outer Space." United Nations–Republic of Korea Workshop on Space Law. Available from http://www.oosa.unvienna.org/SAP/act2003/repkorea/presentations/specialist/ost2/tennen.doc; accessed March 19, 2005.

11. Dinkin, Sam. 2004 (July 12). "Don't Wait for Property Rights." *Space Review.* Available from http://www.thespacereview.com/article/179/1; accessed March 19, 2005.

12. Hongkyun, Shin. 2003. "System of Property Rights in Outer Space." United Nations–Republic of Korea Workshop on Space Law.

13. Wasser, Alan B. 2004. "The Space Settlement Initiative." Space Settlement Institute. Available from http://www.spacesettlement.org/; accessed March 19, 2005.

14. van Ballegoyen, Arjen F. 2000 (January/February). "Ownership of the Moon and Mars?" *Ad Astra.* Available from http://www.space-settlement-institute.org/Articles/archive/BallegoyenOwn.pdf; accessed March 19, 2005.

15. United Nations. 1969. "Vienna Convention on the Law of Treaties." 1155 U.N.T.S. 331. Available from http://www.amanjordan.org/english/un&re/un2.htm; accessed July 1, 2005.

16. White, "Real Property Rights in Outer Space."

17. van Ballegoyen, "Ownership of the Moon and Mars?"

18. Markoff, Marco G. 1970. "Space Resources and the Scope of the Prohibition in Article II of the 1967 Treaty." *Proceedings, 13th Colloquium on the Law of Outer Space.* American Institute of Aeronautics and Astronautics, 81–83.

19. Markoff, "A Further Answer."

20. Williams, Silvia Maureen. 1970. "The Principle of Non-Appropriation Concerning Resources on the Moon and Celestial Bodies." *Proceedings, 13th Colloquium on the Law of Outer Space.* American Institute of Aeronautics and Astronautics, 157–159.

21. Goldman, Nathan C. 1988. *American Space Law: International and Domestic.* Iowa City: Iowa State University Press, 70.

22. Danilenko, Gennady M. 1989. "Outer Space and the Multilateral Treaty-Making Process." *Berkeley Technology Law Journal,* 4, 2. Available from http://www.law.berkeley.edu/journals/btlj/articles/vol4/Danilenko/HTML/text.html; accessed March 19, 2005.

23. White, "Real Property Rights in Outer Space."

24. United Nations Committee on the Peaceful Uses of Outer Space. 1979. U.N. Doc. A/AC.105/PV.203, 22.

25. United Nations. 1982. "United Nations Convention on the Law of the Sea." 1833 U.N.T.S. 3. Available from http://www.un.org/Depts/los/convention_agreements/texts/unclos/unclos_e.pdf; accessed November 29, 2004.

26. Russian Federation. 1993. "Law of the Russian Federation on Space Activity (June 20, 1993)," Art. 16, Para. 4. Available from http://www.jaxa.jp/jda/library/space-law/chapter_4/4-1-2-7/4-1-2-73_e.html; accessed March 18, 2005.

27. Dinkin, "Don't Wait for Property Rights."

28. Markoff, "Space Resources."

29. Gorove, Stephen. 1977. *Studies in Space Law: Its Challenges and Prospects.* Leyden: A. W. Sijthoff, 79.

30. Ibid., 179; and Wadegaonkar, Damodar. 1984, *Orbit of Outer Space Law.* London: Stevens and Sons, 34.

31. White, "Real Property Rights in Outer Space."

32. Csabafi, Imre. 1965. "The UN General Assembly Resolutions on Outer Space as Sources of Space Law." *Proceedings, 8th Colloquium on the Law of Outer Space.* American Institute of Aeronautics and Astronautics, 337–361.

33. Ibid.

33. White, "Real Property Rights in Outer Space."

34. Ibid.

35. Ibid.

36. Katzenbach, Nicholas. 1965, "The Law in Outer Space," in *Space: Its Impact on Man and Society,* ed. Lillian Levy. New York: W. W. Norton & Company.

37. White, "Real Property Rights in Outer Space."

38. Ibid.

39. Ibid.

40. Ibid.

41. Neureiter, Norman P. 2002 (October 12). "Keynote Address to Space Policy Summit." Houston, Texas. Available from http://www.state.gov/g/oes/rls/rm/2002/14540.htm; accessed March 25, 2005.

42. Brill, Kenneth. 2002. "Statement of Ambassador Kenneth Brill, Permanent Representative of the United States of America to the United Nations in Vienna." 41st session of the Legal Subcommittee of the United Nations.

43. Hodgkins, Kenneth. 2002 (October 9). "Statement of Kenneth Hodgkins, U.S. Adviser to the Fifty-Seventh Session of the UN General Assembly, Statement in the Fourth Committee," New York, New York. Available from http://www.state.gov/p/io/rls/rm/2002/14261.htm; accessed March 24, 2005.

44. Wasser, Alan. 1997 (March). "How to Restart a Space Race to the Moon and Mars." *Moon Miners' Manifesto, #103.* Available from http://www.asi.org/adb/06/09/03/02/103/space-race.html; accessed March 19, 2005.

45. Silber, Kenneth. 1998 (November). "A Little Bit of Heaven: Space-Based Commercial Property Development." *Reason.* Available from http://www.findarticles.com/p/articles/mi_m1568/is_n6_v30/ai_21231184; accessed March 19, 2005.

46. Snider, John C. 2000 (May). "SpaceDev Conquers the Universe!" Scifi dimensions.com. Available from http://www.scifidimensions.com/May00/real_tech_spacedev.htm; accessed March 29, 2005.

47. Dinkin, Sam. 2004 (May 10). "Property Rights and Space Commercialization." *Space Review.* Available from http://www.thespacereview.com/article/141/1; accessed March 19, 2005.

48. Dinkin, Sam. 2004 (June 21). "The Dinkin Commission Report (Part 1)." *Space Review.* Available from http://www.thespacereview.com/article/164/1; accessed March 19, 2005.

49. Dinkin, Sam. 2004 (July 26). "Space Privatization: Road to Freedom." *Space Review.* Available from http://www.thespacereview.com/article/193/1; accessed March 19, 2005.

50. Feeney, Tom. 2004 (July 22). "Private Property and a Spacefaring People." Coalition for Property Rights. Available from http://www.proprights.com/newsviews/display_newsletter.cfm?ID=96; accessed March 19, 2005.

51. Danilenko, "Outer Space and Multilateral Treaty-Making."

52. Neureiter, "Keynote Address."

53. United States House of Representatives. 1979. "International Space Activities, 1979." Hearings Before the Subcommittee on Space Science and Applications, Committee on Science and Technology. 96th Congress, 1st Session. 5 and 6 September. Y4.Sci2:96/50, 83; and United States Senate. 1980. "The Moon Treaty." Hearings Before the Subcommittee on Science, Technology, and Space of the Committee on Commerce, Science, and Transportation. 96th Congress, 2nd Session, 28.

54. Christol, Carl Q. 1983. "The *Jus Cogens* Principles and International Space Law," *Proceedings, 26th Colloquium on the Law of Outer Space.* American Institute of Aeronautics and Astronautics, 1; and Goldman, *American Space Law,* 69.

55. United Nations, "Vienna Convention on the Law of Treaties."

56. Jennings, R. Y., and A. Watts .1996. *Oppenheim's International Law,* 9th ed. New York: Longman, 7–8.

57. O'Donnell, Declan J. 2001. "United Societies in Space, Inc." Available from http://www.angelfire.com/space/usis/oldsite.html; accessed 6 May 6, 2005.

58. O'Donnell, Declan J. 1999. "Property Rights and Space Resources Development." Available from http://www.mines.edu/research/srr/ODonnell.pdf; accessed February 1, 2005.

59. Ibid.

60. Ibid.

61. O'Donnell, "United Societies in Space, Inc."

62. Ibid.

63. O'Donnell, "Property Rights and Space Resources."

64. Ibid.

65. Ibid.

66. O'Donnell, Declan J. 2001. "Constitution for the Regency of United Societies in Space." *Space Governance Journal,* 6, 22. Available from http://www.angelfire.com/space/usis/constitution.html; accessed February 22, 2006.

67. Gangale, Thomas, and, Marilyn Dudley-Rowley. 2004. "The Architecture of Time: Design Implications for Extended Space Missions." Society of Automotive Engineers. SAE 2004-01-2533. 34th International Conference on Environmental Systems. Colorado Springs, Colorado. *SAE Transactions: Journal of Aerospace.* Available from http://www.sae.org/servlets/productDetail?PROD_TYP=PAPER&PROD_CD=2004-01-2533; accessed January 21, 2006; Gangale, Thomas, and, Marilyn Dudley-Rowley. 2005. "Issues and Options for a Martian Calendar." *Planetary and Space Science.* 53:1483–1495. Available from http://www.sciencedirect.com/science?_ob=ArticleURL&_udi=B6V6T-4H8FPXR-2&_user=10&_coverDate=10%2F06%2F2005&_rdoc=1&_fmt=summary&_orig=browse&_sort=d&view=c&_acct=C000050221&_version=1&_urlVersion=0&_userid=10&md5=5213bc3d244197d16759c6cd83e2b33a; accessed January 21, 2006; and Gangale, Thomas. 2006. "The Architecture of Time (Part 2): The Darian System for Mars." Society of

Automotive Engineers. 06ICES-2. 36th International Conference on Environmental Systems. Norfolk, Virginia.

68. O'Donnell, "Property Rights and Space Resources."

69. Ibid.

70. O'Donnell, Declan J. 2001. "Proceedings of the Denver Space Constitutional Convention, August 4, 2000 to 2001." Available from http://www.angelfire.com/space/usis/constitution.html; accessed February 22, 2006.

4

Common Heritage in Magnificent Desolation: The Moon Agreement's Tragic Odyssey

THE QUEST FOR CONSENSUS

In the 1970s, the international community negotiated two agreements to manage the natural resources of the international commons (i.e., natural resources in locations that were not subject to national claims of sovereignty). The third United Nations Convention on the Law of the Sea (UNCLOS III) was of more immediate importance than the Moon Agreement because of the perceived accessibility of commercially valuable resources on the ocean floor.[1] However, the international political-economic climate of the time was reflected in the negotiations of both treaties. The world map had changed a great deal since the beginning of the Space Age two decades earlier. The European colonial empires had dismantled themselves in the 1960s, giving rise to dozens of newly independent states. Many of these states were impoverished and saw the root of their underdevelopment in their exploitative economic relationships with the former colonial powers. These economic ties had largely survived the transition to independence; private interests in the former colonial powers owned most of the assets in the newly independent states and continued to extract the profits from those assets. Even long-independent states in Latin America pointed to an enduring pattern of economic dependence enforced by the relationship of the peripheral Third World (characterized by trade in low value-added primary goods) to the industrialized core (characterized by trade in high value-added services and manufactured goods). An informal coordinating mechanism known as the Group of 77 formed, which eventually grew to include more than 120 less-developed countries. This group promulgated the New

International Economic Order (NIEO), a broad program to change the existing rules of the global economy and enable the Third World to catch up with the developed world.[2]

At the same time, however, concern grew over the "carrying capacity" of Earth. Could the environment withstand industrialization on a global scale? Where would the resources come from to sustain billions of people on Earth, and sustain them at the same level of affluence as the few hundred million currently affluent Japanese, Americans, and Western Europeans? An answer to the question concerning "limits to growth" was the "high frontier" and the "deep frontier"—seeking new sources of minerals and energy beyond the sky and beneath the sea.[3] The negotiations for both the UNCLOS III and the Moon Agreement were buffeted by the turbulent diplomatic struggle for jurisdiction over these distant resources and their direct bearing on future economic development. Several Latin American states, including Brazil, Colombia, and Venezuela, regarded the status of natural resources in the Moon Agreement as an important step in the establishment of an NIEO.[4] The phrase "common heritage of mankind" was used in the Declaration of Principles Governing the Sea-Bed and the Ocean Floor, and the Subsoil Thereof, Beyond the Limits of National Jurisdiction to describe the legal character of the natural resources in the oceanic commons.[5] It was then applied to the celestial commons in the negotiation of the Moon Agreement. The Third World wanted one of two things: either the launching states capable of mining the planets to share some of benefits with the less fortunate of the Earth, or the assurance that there would still be extraterrestrial resources available when future launching states emerged in the Third World.

"Common heritage of mankind" is not necessarily the same thing as the "province of all mankind" principle in Article 1 of the Outer Space Treaty. The latter was equated with "for the benefit of all mankind" in the U.S. Senate hearings of that treaty, and the Soviet delegation to COPUOS construed the term to mean that "celestial bodies are available for the undivided and common use of all States on Earth, but are not jointly owned by them."[6] One problem with the term "common heritage of mankind" is that it seems to have substantially different meanings to different people. To some it conjures a highly communistic vision of the proceeds from resource exploitation being monetarily distributed to every human being. Ironically, Soviet law professor R. V. Dekanozov favors the term "international resources" to indicate that they are available for acquisition as property rather than as the province of any individual state.[7] Arthur M. Dula expressed the fear that "common heritage of mankind" would be interpreted as common ownership, and implied that the Soviet Union was behind this alleged limitation on free enterprise in outer space:

> The United States will soon decide whether to sign and ratify the proposed U.N. 'Moon Treaty' designed to control all activities on all celestial bodies in the solar system other than the earth as well as the use of all trajectories to and around them. An analysis of the draft treaty shows that it would create a moratorium on commercial exploitation of the resources of the moon and other celestial bodies until a second, more comprehensive treaty for regulating such activities is concluded, and establish guidelines for this second treaty antithetical to the commercial development of outer space resources by private enterprise. This would allow the USSR and the Third World to decide when to expand commercial uses of outer space or whether to permit them at all. Like the Law of the Sea, which it closely parallels, this Law of Space would limit the basic legal rights that free enterprise will need to work effectively in space. The conclusion is that the proposed treaty is neither necessary, desirable, nor in the best interests of the U.S. and the free world.[8]

Dula interprets the entire history of space treaty negotiations in terms of communism versus capitalism, with the Soviet Union doing its best to disadvantage the United States at every opportunity:

> As introduced by the U.S.S.R. in 1962, the [Outer Space Treaty] forbids free enterprise in space.[9]

The wording here is deceptive. The ratified version of the Outer Space Treaty certainly does not forbid free enterprise in space, as evidenced by all of the free enterprise that has been conducted in space since it has been in force. That an early Soviet draft declaration may have attempted to forbid free enterprise in space is irrelevant as a point of law to the Outer Space Treaty and as a point of law to the Moon Agreement.* The argu-

---

* Also, as noted in the "Declaration of Legal Principles" section in Chapter 2, the Soviet Union abandoned this proposal a year after they submitted it. The Soviets did not want free enterprise, but realized that they would not get agreement on banning it altogether. Their fallback plan was to try limiting free enterprise by requiring states to be legally responsible for all space activity by their nationals. The Soviets were similarly unenthusiastic about intergovernmental organizations having scope for independent activity. Soviet disinclination to repeat the UNCLOS III seabed regime experience in space partial explains why the Moon Agreement ended up being an agreement to come to later agreement on resource activity.

ment that Dula attempts to make is political, not legal, and even its political import is highly suspect:

> The [1971] draft Moon Treaty went through numerous minor changes within COPUOS over the next eight years. At the end of the 1979 COPUOS session consensus was reached on its present language.[10]

Here Dula makes it sound as though the Soviet Union eventually received everything it wanted. Dula was not present at the negotiations, however, whereas U.S. representative to COPUOS S. Neil Hosenball was—year after year. Hosenball's gave testimony on July 29, 1980, before the Senate Commerce, Science, and Transportation Committee's Subcommittee on Science, Technology and Space. He paints a starkly different picture that Dula does:

> The Soviet Union throughout made a whole series of concessions, as I view it. . . .
>
> There was a provision in their draft treaty which our delegation believed could be interpreted to exclude the use of other entities than states or state organizations in exploitation or on exploration. That was subsequently dropped by the Soviet Union.
>
> The Soviet Union wished to limit the treaty to the Moon. They subsequently agreed that the treaty would govern other celestial bodies. The Soviet Union for almost the entire period from 1972 until we finally reached consensus would not agree to common heritage language in the treaty. That one issue, I think, more than anything else delayed a consensus being reached. . . .
>
> If you examine the 1971 Soviet proposal, you will find some elements of the Soviet proposal in the final text, but there is a great deal of difference between what was in the Soviet proposal and what was finally agreed to.[11]

In fact, the 1971 Soviet "Draft Treaty Concerning the Moon" did not even mention the "common heritage of mankind," even though Argentina had proposed that language a year earlier.[12] The record of negotiation shows that the Soviet Union was more jealous of its planetary prerogatives than was the United States (or any capitalist nation for that matter) in that it was not eager for the Third World to acquire the legal authority to dictate the disposition of extraterrestrial resources extracted by Soviet labor, which it considered the property of the Soviet people. P. P. C. Haanappel dismisses Dula's fears of collusion against private enterprise as unfounded.[13] The truth is that the draft proposal that the United States submitted on April 17, 1972, adopted the "common heritage of mankind" phrase, which was consistent with the position that the Nixon administration had taken in 1970 regarding the resources of the seabed.[14] The Soviet

Union continued to oppose the "common heritage of mankind" language in the Moon Agreement until a qualifying clause, "which finds its expression in the provisions of this Agreement and in particular in paragraph 5 of this article," was appended on the last day of negotiation, thus drastically circumscribing its meaning. The Soviet position was that outer space was *res nullius* rather than *res communis* (which is something of a historical irony, given the *Communist* Party's rule of the Soviet Union*), that is, the Moon and other celestial bodies were available for the undivided and common use of all states but not jointly owned by them. No state had a legal claim on extraterrestrial resources that it did not itself obtain. Thus the Soviets adhered to the prior language of the Outer Space Treaty, "the province of all mankind," which was also carried forward in Article 4, paragraph 1 of the Moon Agreement.[15] On the other hand, the documentary evidence shows that Italy, an industrialized, capitalist democracy, was Argentina's staunchest ally in promoting a radical view of *res communis*.[16]

Dula also claims:

> If ratified by the United States, the Moon Treaty's provisions will control the activities of the United States, as well as those of all U.S. citizens and organizations, not only on the Moon, but also on every celestial body in the solar system (other than the Earth) and in the trajectories around and between them. It is hornbook law that any preexisting U.S. law or regulation contravening a ratified treaty is void.[17]

This is certainly true, but it is also hornbook law (a fundamental legal principle) that Dula's last sentence here only holds for self-executing treaties. Most treaties are non-self-executing in the eyes of U.S. law and require federal legislation to execute the provisions of international law in U.S. municipal law. Thus, the ominous warning that "the Moon Treaty's provisions will control the activities of the United States, as well as those of all U.S. citizens and organizations, not only on the Moon, but also on every celestial body in the solar system (other than the Earth) and in the trajectories around and between them" can only succeed in frightening hornbook lawyers. It should be noted that whether a treaty is or is not self-executing does not determine the extent of U.S. government obligation to respect the treaty stipulations; rather, the question of self-execution is whether private entities or individuals can invoke a treaty in U.S. courts immediately or must wait for and be bound by the terms of implementing

---

* In fact, the Soviet Union's objection was motivated in part by solid Marxist principles, which deny the right to inherit property. It argued from 1972 to 1979 against the inclusion of the "common heritage" principle on the basis that it could have no meaning, since the human community cannot inherit what is not owned by any other entity.

legislation. There is a well-established international legal principle that a state may not invoke its own domestic law as an excuse for violating international law This principle would answer arguments that the lack of domestic implementing legislation means the government cannot undertake actions to fulfill U.S. treaty obligations. The operating assumption in international law, which is sometimes in tension with the separation and balance of powers in the United States, is that a government controls its own legislative processes and accomplishes whatever is needed domestically to make treaties work. Nevertheless, there may be some latitude in a state party's determining how to execute an international agreement in its municipal law.

Dula further asserts that

> [t]he Moon Treaty is vague, lengthy, and complex. . . . The language of its most important articles closely parallels language in the 1970 U.N. Resolution on the Deep Seabed and the draft Law of the Sea Convention.[18]

"Lengthy and complex" compared to what? Compared, for instance, to the Law of the Sea Convention itself? Surely not. The basic UNCLOS III document runs 208 pages (76,000 words), not including subsequent amendments, whereas the Moon Agreement barely covers 9 pages (just under 3,600 words). And which articles in the Law of the Sea Convention "closely parallel" the important articles of the Moon Agreement? If such close parallels exist, a comparative analysis, citing chapter and verse, should be easily provided. See chapter two's "Legal Principles" section as an example this type of analysis. In a 33-page law journal article, Dula does not provide such an analysis. No other opponent of the Moon Agreement who claims it contains parallels with the Law of the Sea Convention provides such a comparison. One must wonder why.

Dula complains, as do others, that the Moon Agreement does not define the terms "celestial body" and "natural resources"; however, he makes no point from this observation. One could similarly observe that the Outer Space Treaty does not define the term "outer space." For that matter, Part 1 of the Law of the Sea Convention defines several terms, but "sea" is not one of them. Are such definitions necessary? Every time one boards an aircraft, one is traveling in an environment that has no international legal definition. Do we not understand what the air, the sea, and the Moon are unless we agree on tortuous legal definitions?*

Reasonable person standard

---

* Thomas Gangale argues that existing treaty language implies that space objects are objects that are either in Earth orbit or beyond, thereby constituting a *de facto* functional definition of outer space.[19] Suborbital objects continue to be a gray area in air and space law.

Regarding the validity of Dula's allegation of a moratorium on commercial exploitation of the Moon, see chapter five, "A Mining Moratorium."

The term "common heritage of mankind" was being used in the UNCLOS III then in negotiation, and that convention contained specific implementation criteria. The question thus arose whether "common heritage of mankind" meant the same thing in both treaties, and whether acceptance of the term in the Moon Agreement would lead to a lunar resources exploitation regime similar in scope to the one described in the UNCLOS III. Edward R. Finch and Amanda Lee Moore, authors of *Astro Business: A Guide to Commerce and Law of Outer Space*, state:

> The major controversy over the Moon Treaty concerns principally Article II(1) and (5). Paragraph 1 states that the moon and its natural resources are the "common heritage of mankind." This term is without specific, agreed definition, as evidenced by the remaining text of that paragraph, which says that the term "finds its expression in the provisions of this Agreement and in particular in paragraph 5 of this article."[20]

The Moon Agreement achieved separation from the Law of the Sea by indicating that the term "common heritage of mankind . . . finds its expression in the provisions of this Agreement." In other words, for the purpose of the Moon Agreement, the "common heritage of mankind" only meant what the agreement itself meant, and that meaning was found in the text of the agreement itself and, to some extent, in the negotiating history of the agreement. Fielding a question from the Senate Space Subcommittee, Hosenball writes that the Soviets

> appear to feel that interpretation of the common heritage concept cannot be linked to precedents outside of the Moon Treaty, such as the Law of the Sea.[21]

In interviewing COPUOS delegates, Finch and Moore gained the general impression that "the phrase is in essence a continuation of the very general concept from the 1967 Outer Space Treaty of space as the 'common province of mankind'" with an attempt to move into language more commonly used in international law. They also observed that "no two delegations have said the term means the same thing at any given time." They further state:

> The essence of the concept of common heritage in the context of the Moon Treaty is said to lie in the existence of common, that is, equally shared, rights to explore and use the moon and its natural resources. It does not, however, connote specific implementing criteria or procedures.[22]

Because they weren't thinking far enough ahead

Eilene M. Galloway of the International Institute of Space Law (IISL) finds the same limitation of the meaning of the "common heritage of mankind" term:

> Paragraph 1 . . . provides that "the moon . . . and its natural resources are the common heritage of mankind . . ." a general principle implying that every person in the world has a stake in this heritage; but the paragraph continues with specifics: "which finds its expression in the provisions of this Agreement and in particular in paragraph 5 of this article." Thus the general principle is to be limited in its implementation.[23]

Since the "common heritage of mankind" term had no agreed definition, one commentator concludes that the term was "purely declaratory . . . and open to all interpretations."[24] If it has no agreed definition, it has no legal force, nor is a state obliged to act against its interests on the basis of the term.

Even the man who introduced the term into space law in 1970, Aldo Cocca, the Argentine representative in COPUOS, reflects in a letter to Galloway:

> [I]t is rather dangerous to crystallize in a definition the principle involved in a concept which is just being born in the new domain of Space Law, such as the "common heritage of mankind," as it was established in the Moon Agreement. . . . I daresay it is not a matter of definition; I feel it must be the outcome of the implementation of the guidelines set forth in the agreement.[25]

THE DONE DEAL . . . UNDONE

There may have been some degree of rumormongering inside the Washington, DC, beltway while the U.S. State Department was preparing to sign the agreement and present it to the Senate for ratification. Charles Chukwuma Okolie cites Harry H. Almond, senior attorney-advisor, International Law and Affairs Division, U.S. Department of Defense, as writing that Hosenball's statements regarding the agreement did not reflect a consensus of opinion within the domestic politics of the United States, and that members of the Senate Committee on Foreign Relations were displeased with the agreement. It should be noted first that Hosenball's statements in COPUOS and its Legal Subcommittee are part of the official record of negotiation of the agreement, which under international law constitute a source for interpretation of the agreement that is secondary only to the text of the agreement itself.

Whatever anyone else on the U.S. domestic political scene might think about Hosenball's statements in the official record is considerably less relevant. Second:

> Mr. Almond's sources were not very accurate. It appears that he derived his impressions from a number of "speakers at several conferences in the Washington area who believe that the clauses in Article XI—those declaring the 'Moon and its natural resources' to be the 'common heritage of mankind' and those declaring that 'the Moon is not subject to national appropriation by any claim of sovereignty, by any means of use of occupation, or by any other means'—effectively bar exploitation by private enterprise." We believe that Mr. Almond has misunderstood or misinterpreted his sources and thus made an erroneous analysis of the perception of the Moon Treaty in the legislative circles, as well as of the facts surrounding the 1979 Draft Moon Treaty. We would like to remind him that, after all, the U.S. is the developer of the concept of common heritage of mankind; at least the use of the term was first employed by the U.S. in its official answer to Premier Nikita Khruschov [sic] in 1957 after the Soviets had successfully launched the first artificial satellite into space.[26]

Whatever might have been the perception of the members of the Foreign Relations Committee prior to Almond's remarks in January 1980, an effort to sink the Moon Agreement was already in motion. Shortly after the agreement had been opened for signature in 1979, a small space enthusiast organization called the L-5 Society mobilized political opposition to the agreement. The society had formed four years earlier with the goal of constructing self-sustaining colony at the Earth-Moon Lagrange point L-5,* using lunar materials and O'Neill's ideas as a basis (see Figure 4.1). The society saw the "common heritage" principles of the Moon Agreement as a barrier to the use of lunar materials for the L-5 colony and to the development of outer space in general. For some reason, L-5 members also feared that the agreement would limit the freedom of individuals and groups in space.

---

* The Lagrange points are the stationary solutions of the circular restricted three-body problem. Given two massive bodies in circular orbits around their common center of mass, there are five positions in space where a third body, of comparatively negligible mass, could be placed which would then maintain its position relative to the two massive bodies. As seen in a frame of reference that rotates with the same period as the two co-orbiting bodies, the gravitational fields of the two massive bodies are in balance at the Lagrange points, allowing the third body to be stationary with respect to the first two bodies. L_1, L_2, and L_3 are unstable, but L_4 and L_5 are stable.

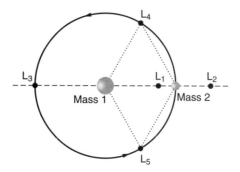

Figure 4.1: Lagrange Points
Source: NASA.

According to Galloway, when COPUOS reported the final text of the agreement to the General Assembly in July 1979:

> The first reaction was surprise because some interested people had assumed that the Moon Treaty would never achieve consensus; others were alerted for the time to United Nations international space activities. The issues that have arisen for discussion during the past year are different from those which were ultimately reconciled within the United Nations. A regrettable aspect of the debate over issues is that much of the information published and distributed on this subject contains factual errors, short changes presentation of "the whole truth," and generally suffers from lack of an objective research base.
>
> Without doing any research, some writers assumed that the United Nations had put something over on the United States, whereas there is a clear record that the United States achieved in the Moon Treaty all the major points of its policy as pursued by the Nixon, Ford and Carter administrations. In 1972 the United States proposed formally the concepts of the common heritage of mankind and an eventual international regime. . . . A strange element in the present situation is that the opponents of some of these proposals waited for eight years before voicing their objections.
>
> I have worked with research materials on public affairs for several decades and have never before encountered a subject about which there has been so much misinformation and misinterpretation as that on the proposed Moon Agreement.
>
> . . . [M]uch of the material produced during the past year has been emotional, intemperate, speculative, one-sided and often inaccurate.[27]

The opening shot in the battle of the Moon Agreement occurred when the previous L-5 Society president H. Keith Henson, husband of then-president Carolyn Henson, declared with considerable intemperance:[28]

> On the Fourth of July 1979 the space colonists went to war with the United Nations of Earth. . . . The treaty makes no provisions for the civil rights of those who go into space. In fact, it authorizes warrantless searches. . . . The treaty makes about as much sense as fish setting the conditions under which amphibians could colonize the land.[29] *Damn rebels*

Other space enthusiast groups joined in condemning the agreement and the people who negotiated it. The Space Futures Society, allied with the L-5 Society, is particularly vitriolic in its letter, written by its director of public information, Michael Calabrese, to the Senate Space Subcommittee:

> [N]o agency of office of our government has a definitive understanding of this treaty . . .
> [T]he treaty as it presently exists, not only operates against the best interests of the United States, it serves the interests of the Soviet Union.
> Ever since the publication of Lenin's writings before the 1917 revolution, it has been the strategy of the Soviet Union to separate the West from its energy and industrial resources. . . .
> The federal government of the United States is almost totally ignorant of the value of space, its potential for development as an industrial base or its economic benefits. There is a basic lack of understanding in terms of just what the American ability in space is and what it means to this nation and the free world. It is this condition in the thinking of the American government that creates much of the benefits for the Soviet Union in this treaty.[30]

A hawkish view

Calabrese heaps abuse on the Carter administration in apparent ignorance of the fact that the Moon Agreement was already near its final form when Jimmy Carter took office in January 1977 and that the final two years of negotiation centered on the Soviet Union's obstinacy over the "common heritage of mankind" principle.

THE OCEAN OF STORMS

In August 1979, Carolyn Henson contacted Leigh Ratiner, a Washington lawyer and lobbyist who was then engaged in lobbying against provisions in the draft UNCLOS III (which was still being negotiated at that time), pertaining to a proposed deep seabed mining regime having some parallels to the mining regime envisioned in the Moon Agreement. Ratiner had directed the Interior Department's Office of Ocean Affairs in the Ford administration, and was later instrumental in crafting the Reagan administration's position on the UNCLOS III seabed mining issue. Since the issues in the two treaties were similar, at least on a superficial level, the normative objections were seen to be similar and mutually reinforcing. Ratiner told Henson that the Moon Agreement could be defeated for about $100,000, which was certainly affecting the future of space law and policy on the cheap, but after all, the Moon Agreement opposition was really riding on the momentum of the much larger issue of the UNCLOS III seabed mining issue. In September 1979 the L-5 Society board voted to oppose the agreement, despite its support by board member and international lawyer Edward R. Finch, and hired Ratiner. Society members organized a letter-writing and telephoning campaign targeting key members of the U.S. Senate in an effort that continued into 1980. It helped that Ratiner was a friend of Representative John Breaux (D-LA), chair of the House Subcommittee on Fisheries, Wildlife, and Environment, who invited him to testify against the agreement. Ratiner gained enough credibility from this appearance to approach staffers of the Senate Foreign Relations Committee, which would consider the agreement once the administration signed it. He prepared a draft letter from the committee to Secretary of State Cyrus Vance, which formed the basis of a letter signed on October 30, 1979, by committee chair Frank Church (D-ID) and ranking minority member Jacob Javits (R-NY), urging that the United States not sign the agreement:

> After a decade of negotiation at the Law of the Sea Conference, the set of draft treaty articles now before the Conference sets forth an interpretation of the "common heritage" which does not conform to the national interest of the United States or of other countries with free enterprise/free market economies, particularly as they relate to such matters as production limitations, technology transfer, dispute settlement and competition with the proposed international "Enterprise."[31]

The senators objected to the Moon Agreement on the basis of the perceived similarities between it and the Law of the Sea. What similarities? The Moon Agreement contains no language regarding "production limitations, technology transfer, dispute settlement and competition with the proposed international 'Enterprise.'"* Nevertheless, Representative Breaux, who opposed the seabed mining language in the UNCLOS III,

---

* The "Enterprise" was to be an entity of the United Nations that would mine the seabed and allegedly distribute its profits internationally according to a socialist scheme, allegedly being given preferential treatment by the International Seabed Authority and thus driving out competition from privately owned corporations.

sent a similar letter to Secretary Vance. All of this occurred as the agreement came up for action by the UN General Assembly, where it passed with the support of the United States. At that point, however, Vance suspended action on the agreement. The State Department formed an interagency group to study the agreement, and it was never signed.[32] Ratiner achieved what he had set out to do, "to kill this treaty dead."[33] Ever since, the American space enthusiast community has pointed to the defeat of the Moon Agreement in the United States as one of its greatest triumphs. However, historian Michael A. G. Michaud reflects:

> It seems clear that members of Congress (and Ratiner) opposed the treaty primarily for Law of the Sea reasons and not because they were advocates of space development. Their major concern was to prevent restrictions on seabed mining. Support for the treaty was weak and unorganized outside a small group of international lawyers involved in its negotiation and some sympathetic academics; it was relatively easy to kill. What L-5 should have learned from this experience is the value of weak interest groups having more powerful allies. "L-5 was the tail on a very large dog," comments [space colonization advocate] Thomas Heppenheimer.[34]

Looking outside the self-congratulatory hype of the space enthusiast community, one sees that larger issues defeated the Moon Agreement. A broader perspective of the international political-economic issues of the day reveals that the L-5 Society provided useful foot soldiers and a few platoon leaders for the command of a Beltway insider in the opening battle of a campaign whose ultimate objective was the defeat of the UNCLOS III seabed mining regime. The smoking gun in this plot is the testimony of Marne A. Dubs, chairman of the Committee on Undersea Mineral Resources of the American Mining Congress, and vice-president of the mining firm, Kennecott Development Corporation:

> I believe if we sign that treaty, for example, that it would seriously interfere with our remaining arguments on the Law of the Sea Conference as far as negotiating a more acceptable agreement.[35]

Ronald F. Stowe, chairman of the Aerospace Law Committee, Section of International Law, American Bar Association (ABA), testifying before the Senate Space Subcommittee regarding the Moon Agreement and its purposes, states:

> Proponents of nonratification appear to justify their recommendation principally with the argument that, in their view, the United States has been taken to the cleaners in the Law of the Sea negotiations, and we should therefore refuse to participate in any comparable exercises.

In my view, that suggests an unacceptable futility and self-fulfilling anticipation of defeat.[36]

Nevertheless, the Ratiner strategy worked brilliantly. Even the ABA Section of Natural Resources Law fell for it, which is hardly surprising, given that its report showed detailed knowledge of the Law of the Sea negotiations but no knowledge whatsoever of the Moon Agreement's negotiating history.[37] The upshot was that the ABA, split in its stance, was unable to take a position one way or another. Among the aerospace professional organizations, the American Astronautical Society declared its opposition to the agreement, while the American Institute of Aeronautics and Astronautics waffled.[38]

WHAT LIES BENEATH

It is odd that there was a great deal of interest in commercial seabed mining in the 1970s, just as the UNCLOS III and the Moon Agreement were being negotiated, yet within a few years of the defeat of both treaties in the United States, the issue quietly faded into oblivion. Herein lies another smoking gun in the death of the Moon Agreement, but it was an accidental shooting. The events that resulted in the death of the agreement began in 1968, years before its gestation began in COPUOS meetings.

In April 1968, U.S. surveillance assets observed Soviet Pacific Fleet surface vessels and aircraft conducting an unusual surge deployment in the North Pacific. U.S. Naval Intelligence interpreted the activity as a possible reaction to the loss of a Soviet submarine. Soviet surface ship searches were centered on an area in which Soviet strategic ballistic missile submarines were known to operate. The American Sea Spider hydrophone network in the northern Pacific reviewed its recordings in the hopes of detecting an implosion related to such a loss. Correlating the data from five listening stations from Alaska to California, the U.S. Navy determined that an implosion event had occurred on March 8, 1968, near 40°N, 180°E. The diesel-powered *K-129* had gone down with all hands.

Despite weeks of effort, the Soviets were unable to locate *K-129*, and Soviet Pacific Fleet operations gradually returned to a normal level. The U.S. Navy, however, was just getting warmed up. In July 1968, it launched Operation Sand Dollar by deploying *USS Halibut* (SSN-587) from Pearl Harbor to the calculated site of the wrecked *K-129*. The mission's objective was to find and photograph *K-129*. *Halibut* was the only submarine then in U.S. inventory that was specially equipped to use deep submergence search equipment. Finding *K-129* was far from a slam dunk. The coastal hydrophone data was only able to confine the search area to 3,100 square kilometers (1,200 sq mi), and the wreck located over 4.8 kilometers (3 mi) down. With good equipment, good training, and more than a little good luck, *Halibut* spotted the wreck after only three weeks. In comparison, that same year, it took five months to locate the wreck of *USS Scorpion* (SSN-589) in the Atlantic Ocean.

In 1970, based on extensive imagery that *Halibut* obtained of *K-129*, the United States developed a plan to recover a section of *K-129*, which had broken apart on impact, and study its Soviet nuclear missile technology, as well as possibly recover cryptographic materials. President Richard M. Nixon gave a green light to the plan, and the Central Intelligence Agency got to work on what would come to be known as Project Jennifer. The CIA contacted eccentric billionaire Howard Hughes, whose companies had numerous classified contracts with the U.S. military, and asked him to design and build a ship to raise the sunken Soviet submarine from the ocean floor. Its resting place at a depth of over 4,900 meters (16,000 ft) was well beyond the depth of any ship salvage operation previously attempted. On November 1, 1972, Hughes began work on the 63,000 ton *Glomar Explorer.* The cover story for the project was that *Glomar Explorer* was being constructed for the Hughes-owned Summa Corporation to mine for underwater manganese nodules.

The business end of *Glomar Explorer* was a large mechanical claw, built by the Lockheed Corporation and designed to be lowered down to the ocean floor where it would grasp the target section of the submarine and lift it to the open midsection of the ship, called the Moon Pool by its crew. The outer doors of the Moon Pool would then close to form a floor for the salvaged section. This allowed for the entire salvage process to take place underwater, out of view by anyone who might take an interest in *Glomar Explorer*'s operations.

By early 1974, *New York Times* reporter Seymour Hersh had already taken an interest, before *Glomar Explorer* could set out on its mission, and he planned to publish a story on Project Jennifer. The U.S. government gave a convincing argument to delay publication of Hersh's story, pleading that exposure at that time, while the project was ongoing, would have caused an international incident. Of course, even without the leak, it would have taken the Soviets little time to understand *Glomar Explorer*'s real purpose in prospecting the ocean floor in the area of the lost *K-129*, but the United States would have had some deniability, however implausible.

On June 5, 1974, Howard Hughes's headquarters in Los Angeles, California, was burgled, and secret documents about the operation were taken. Ten days earlier, William Turner, Special Assistant United States

Attorney for the District of Nevada, had subpoenaed these documents for a federal grand jury in Nevada conducting an investigation of Hughes and his takeover of Air West Airline. There was also speculation of possible connections between the Los Angeles break-in and the burglary of Democratic Party offices at the Watergate complex in Washington, DC, in June 1972. The documents stolen from Hughes's offices were never recovered; however, there have been rumors that the theft of the documents was part of a cover-up related to political corruption documented by Hughes over several decades.

On June 19, 1974, two weeks after the Los Angeles burglary, *Glomar Explorer* set out from Long Beach, California, arriving at the salvage site on July 4. During the following weeks, at least two Soviet spy ships visited the work site.

Hughes kept a written copy of all his actions and directions, including detailed records of his dealings with political figures and government agencies. These papers would become the foundation of the story on Project Jennifer in the *Los Angeles Times* on February 7, 1975, six months after *Glomar Explorer*'s operations at the *K-129* wreck site. After the story appeared, the CIA attempted to persuade news organizations to refrain from publish further stories on the project, but by March 1975, numerous news stories had linked *Glomar Explorer* to the secret U.S. government operation. The *New York Times* eventually published its account after the *Los Angeles Times* story, and included a chronicle of the story's circuitous route to publication.

According to published reports, on August 12, 1974, *Glomar Explorer*'s claw failed while the section of *K-129* was over halfway to the surface, and the already damaged section broke apart, with all but the forward 12 meters (38 ft.) of the bow section sinking into the depths forever. Officially, the recovered section contained neither the nuclear missiles nor the cryptographic equipment that the U.S. intelligence community sought. Another cover story? Perhaps. In the intelligence community, successes usually go unsung.

The U.S. intelligence community has a term to describe the unintended consequences of covert operations; they call it "blowback." Project Jennifer began to blow in a most unexpected direction, eventually reaching the Moon. Manganese nodules, the recovery for which *Glomar Explorer* was supposed to have been built, were not news. Such nodules had been discovered a century earlier, and some had been recovered by the British research vessel *HMS Challenger*.

By the 1960s, prospective ocean miners talked of scanning seas for nodules rich in the most valued metals. Some did it occasionally. The field's legitimacy took a leap in the early 1970s when Howard Hughes built the *Glomar Explorer,* a 618-foot ship advertised as being constructed to mine the manganese nodules.

In 1974, as the *Glomar Explorer* began its secretive maiden voyage in the Pacific, some 5,000 delegates and observers from 48 nations converged on Caracas, Venezuela, for United Nations deliberations to try to hammer out a sweeping treaty on the Law of the Sea. Understandably, much of the talk focused on the ship and much treaty language was born amid visions of the world's impoverished nations becoming rich by virtue of seabed mining.

The vision persisted even after disclosures in 1975 and 1976 that the *Glomar Explorer*'s mineral work was a fiction meant to hide a Central Intelligence Agency effort to raise a sunken Soviet submarine from deep Pacific waters.[39] *It's a cover, and a great idea!*

Although the cover was blown, the cover story began to take on a monstrous life of its own. After all, just because Hughes got the U.S. government to cough up $350 million for *Glomar Explorer* did not mean that he had only government uses in mind for it. Perhaps he had other, commercial, uses in mind as well. The cover story fueled a perfect storm of corporate acquisitiveness and Third World aspirations that tore apart two treaties. Almost half a billion dollars was invested in identifying potential deposits and in research and development of technology for mining and processing nodules. These initial undertakings were carried out primarily by four multinational consortia composed of companies from the United States, Canada, the United Kingdom, the Federal Republic of Germany, Belgium, the Netherlands, Italy, Japan, and two groups of private companies and agencies from France and Japan. There were also three publicly sponsored entities from the Soviet Union, India, and China. In the mid-1970s, a $70-million international joint venture succeeded in collecting multi-ton quantities of manganese nodules from the abyssal plains of the eastern equatorial Pacific Ocean. Significant quantities of nickel as well as copper and cobalt were subsequently extracted.

In the end, not only was the Moon Agreement never signed by the United States, but when the UNCLOS III was eventually opened for signature in 1982, the United States did not sign it either. The Agreement relating to the Implementation of Part XI of the Convention on the Law of the Sea, meant to address concerns over the proposed seabed mining regime, entered into force in 1996, yet still the United States has not signed.[40]

And still there are no large-scale commercial operations to mine nodules—manganese or any other kind—from the ocean floor. The technology developed during the course of the pilot project was never taken to commercial scale because of an overproduction of nickel by conventional methods in the 1980s and 1990s. The estimated $3.5-billion investment (the ultimate sunk cost) to implement commercialization was an additional factor is the demise of the business case for seabed mining.

A WANING MOON

Remarkable though the L-5 Society's effort was, it must also be noted that no launching state has ever ratified the Moon Agreement, and only two launching states—France and India—have signed it (see Table 2.4). Certainly Ratiner and the L-5 Society effectively leveraged their influence inside the Beltway, but it is highly doubtful that they can take credit for what happened (or did not happen) in other national capitals. What else might account for this global lack of enthusiasm for the agreement?

When work on the Moon Agreement began in 1970

> there seemed to be a pressing need for space law because of the active Moon exploration programs of the United States and the U.S.S.R. during the 1960s, and it could not be foreseen that such programs would diminish during the 1970s. In the case of all other functional space programs, law and technology developed in tandem or not very far apart and this was an essential element in our success.[41]

In fact, by 1970, the United States had already slammed the door on a sustained, manned lunar exploration program by capping Saturn V production at 15 flight items (the last two were never even flown). Plans for an American-manned landing on Mars in the early 1980s had also been dropped. Yet, the Moon Agreement was proposed, and works on it dragged on, year after year. By 1979, it had been seven years since *Apollo 17*, the last lunar expedition, and the last Saturn V launch vehicles were on static display for tourists. The Soviet Union had abandoned its L3 manned lunar landing program in 1972, and development of its N1 launch vehicle in 1974. What had appeared too many to be so close when work began on the agreement now appeared to be decades in the future. According to Gennady M. Danilenko, "Many states failed to ratify the Moon Treaty because they felt that it was premature."[42] More likely, its time had come and gone, and its second coming looked to be a long way off.

During hearings on the agreement, Harrison H. Schmitt (R-NM), the only person to walk on the Moon and sit in the United States Senate, questioned its relevance:

> There is no space policy that relates to this treaty. There is none. What is the frame work for this analysis? That is what I don't understand. I am not

necessarily opposed to the treaty. I would like to know where the administration is coming from on this and what kind of recommendations we might expect to have here so we can better prepare to receive them.[43]

Stanley B. Rosenfield of the New England School of Law, who based his opposition to the agreement on his interpretation of Article 11, states:

> The heart of the proposed Treaty is the Art. XI provisions relating to the exploitation of the resources of the moon and other celestial bodies. Without this provision the Treaty adds little to the treaties already in force. . . . The conclusion seems inevitable that unless Art. XI is satisfactory, there is no justification for this treaty.
>
> It is not likely that it will be possible to exploit such resources for some years into the future. It may be desirable to have a treaty in place when it becomes feasible to exploit the resources of outer space. However, there is no value in rushing through an agreement, the terms of which are ambiguous and in which the most important terms are left to future agreement.[44]

Rosenfield's is a curiously elliptical argument: there is no point in rushing into an agreement to reach further agreement in the future.

Danilenko suggests that the underlying problem with the Moon Agreement was the process that produced it. First, the broadly multilateral scope of the negotiations did not produce a consensus on the meaning of the agreement itself. Second, Danilenko points to the inherent dangers in anticipatory lawmaking.

With respect to the "common heritage of mankind" phrase (Article 11, paragraph 1) and the provision to undertake the establishment of a governing regime (Article11, paragraph 5), Danilenko notes, "The radically opposed interpretations of these provisions indicate the absence of common intent." These "radically opposed interpretations" suggest that the process that gave rise to the agreement was flawed, as COPUOS members should have hammered out consensus on the controversial passages prior to opening the agreement for signature. Furthermore, that there are such "radically opposed interpretations" suggests that the agreement itself is flawed if it cannot "be interpreted in good faith in accordance with the ordinary meaning to be given to the terms of the agreement in their context and in light of its object and purpose."[45]

Both the 1967 Outer Space Treaty and the 1963 Treaty Banning Nuclear Weapon Tests in the Atmosphere, in Outer Space and Under Water (commonly called the Partial Nuclear Test Ban Treaty), which contains some language pertaining to outer space, were the product of negotiations between the three major nuclear powers and space powers of the time: the United Kingdom, the United States, and the Soviet Union.[46] Thus the precedent was set for the most directly interested states establishing prin-ciples and instruments of space law. The failure of the broadly multilat-eral 1979 Moon Agreement, and the lack of any new treaty in the nearly three decades since, gives some weight to the argument that any future agreement must at least have the blessing of the space powers, if not be negotiated and concluded among them exclusively.

However, there are several weaknesses in Danilenko's analysis. To begin with, the negotiation record clearly shows that consensus finally was achieved among the negotiators themselves regarding the legal meaning of the Moon Agreement; that onlookers interpreted the agree-ment according to their own political agendas is something else entirely. Although Danilenko references Article 31 of the Law of Treaties, he fails to consider Article 32, which requires the negotiation record to be taken into account where the means of interpretation specified in Article 31 lead to a conclusion that is either ambiguous or ridiculous. Second, in the broadly multilateral negotiation of the Moon Agreement, the main obsta-cle to consensus was one of the launching states: the Soviet Union. The other major launching state, the United States, agreed early on to the "common heritage of mankind" language, whereas the Soviet Union blocked consensus until 1979 when it accepted compromise language offered by Brazil, as recounted by Hosenball:

> The Soviet Union, for almost the entire period from 1972 until we finally reached consensus, would not agree to "common heritage" language in the treaty. That one issue, I think, more than anything else, delayed a consensus being reached.[47]

This disagreement would have occurred without the involvement of the nonlaunching states in the negotiation of the Moon Agreement, although perhaps it would have been resolved more expeditiously.

Danilenko's second point, regarding anticipatory lawmaking, is on firmer ground. He suggests that anticipatory lawmaking is advantageous to some states:

> [F]rom a political-legal perspective, the anticipatory approach provides states lacking space capabilities better opportunities for an increased role in law-making. Furthermore, the anticipatory approach prevents unfavorable developments in actual practice which may be relied upon by space powers in order to establish effective patterns of behavior reflecting their prefer-ences. In view of this, it is not surprising that, at the official level, the major proponents of early negotiations on space issues are the developing coun-tries who feel that preventative regulation enables them to exert a greater influence on the law-making process.[48]

Nevertheless, there are risks to anticipatory lawmaking:

International space law is based on anticipatory regulation, which produces rules to govern topics that might arise only in the future. . . . The Moon Treaty was negotiated at a time when the activities of states in the exploration and exploitation of the natural resources of the moon were very limited. . . . Subsequent experience indicates that anticipatory regulation may be less appropriate in the formulation of detailed policies regarding complex technical and economic issues.

. . . while anticipatory regulation may be useful for the establishment of a broad legal framework for future space activities, it is dangerous to rely on it too heavily in cases which require detailed regulation of complex technical or economic issues. Early negotiations are usually carried out without substantial knowledge about the subject-matter under discussion. As a result, the law-makers are forced to conduct negotiations based on a number of assumptions about future technological developments, trends in practice and resulting national interests.

The tension between the pressure for anticipatory normative solutions and the dangers of premature regulation became particularly evident in the course of negotiations relating to the legal regime governing the exploitation of the natural resources of the moon. The majority of negotiating states supported the idea of an early normative response to future problems. Other countries, including those specially affected, tried to point out that, at the current stage of development of exploration of the moon, there were no material prerequisites for the detailed regulation of the relevant issues.[49]

The Soviet representative in the Legal Subcommittee of COPUOS emphasized that only

practical experience in the use of the resources of celestial bodies would make it possible to formulate well-founded normative provisions to regulate that aspect of space activity. Otherwise, there [is] a danger that legal norms lacking any practical value might be adopted, norms that would have no relationship to the real tasks and trends of moon exploration and would therefore hamper rather than stimulate that activity, thus having a retrogressive effect.[50]

NASA administrator Robert A. Frosch philosophized:

I am not, as a personal matter, very partial to inventing legalizations in which no one has any experience. That seems to me to be an interesting theoretical exercise, but it is likely that when one arrives at the situation, it is a different situation than was invented beforehand, and may or may not be applicable.

So, I guess I would describe my personal view in the matter as "massive indifference."

I am considerably more concerned with the question of whether we go ahead to develop the technological capabilities and the knowledge to do some exploitation, and if that develops in such a way that we need a new kind of legal regime. Then, it would be useful to negotiate such a thing at a time when there is some knowledge to base it on. I recognize there is a school of thought that opposes my view and says the only time you can negotiate a sensible treaty is when nobody has yet developed a personal interest in it. That is the point at which people can be either neutral enough or, if you like, indifferent enough so they will make the necessary compromises to get a treaty. As soon as somebody has an advantage or a particular interest, then it gets very hard to negotiate treaty law. So you ought to get at it before that happens. My problem is that we seem to be getting at it before anybody understands what they are talking about. That may be all right, I do not know.[51]

The point that Frosch overlooks is that the Moon Agreement merely provides for the possibility of future negotiation on the details of a regime, which by definition would begin "as such exploitation is about to be come feasible," that is, when everybody "understands what they are talking about."

In any case, in holding out for a severe circumscription of the meaning of "common heritage," the Soviet Union ran out the clock on the Moon Agreement. However, a quarter century after the battle of the Moon Agreement, as the United States contemplates a return to the Moon in the next decade, this time with the intent of a sustained program that may possibly lead to commercial development, the question remains: was the defeat of the agreement a good thing? Also, regardless of its merits vis-à-vis the trends in the U.S. and Soviet manned space programs in the early 1980s, a new question arises: does it now make sense to reconsider the agreement?

NOTES

1. United Nations. 1982. "United Nations Convention on the Law of the Sea." 1833 U.N.T.S. 3. Available from http://www.un.org/Depts/los/convention_agreements/texts/unclos/unclos_e.pdf; accessed November 29, 2004.

2. United Nations. 1974 (May 1). "Declaration on the Establishment of a New International Economic Order." General Assembly Resolution 3201 (S-VI). Available from hhttp://www.un-documents.net/s6r3201.htm; accessed October 20, 2008.

3. Meadows, Donella H., Dennis L. Meadows, Jorgen Randers, and William W. Behrens III. 1972. "Limits to Growth: A Report to the Club of Rome." Abstracted by Eduard Pestel. Available from http://www.clubofrome.org/docs/limits.rtf; accessed December 24, 2005; and O'Neill, Gerard K. 1977. *The High Frontier: Human Colonies in Space.* New York: Morrow.

4. United Nations Committee on the Peaceful Uses of Outer Space. 1977. U.N. Doc. A/AC.105/PV.171, 68; United Nations Committee on the Peaceful Uses of Outer Space. 1977. U.N. Doc. A/AC.105/PV.172, 26; and United Nations Committee on the Peaceful Uses of Outer Space. 1978. U.N. Doc. A/AC.105/C.2 SR.291, 6.

5. United Nations. 1971. "Declaration of Principles Governing the Sea-Bed and the Ocean Floor, and the Subsoil Thereof, Beyond the Limits of National Jurisdiction." A/RES/2749 (XXV). Available from http://daccess-ods.un.org/access.nsf/Get?Open&DS=A/RES/2749(XXV)&Lang=E&Area=RESOLUTION; accessed February 22, 2006.

6. United States Senate. 1967. "Treaty on Outer Space." Hearings Before the Committee on Foreign Relations. 90th Congress, 1st Session; and United Nations Committee on the Peaceful Uses of Outer Space. 1977. U.N. Doc. A/AC.105/196, Annex 1.

7. Dekanozov, R. V. 1980. "Juridicial Nature and Status of the Resources of the Moon and Other Celestial Bodies." *Proceedings, 23rd Colloquium on the Law of Outer Space.* American Institute of Aeronautics and Astronautics, 80-SL-03, 5–8.

8. Dula, Arthur M. 1979. "Free Enterprise and the Proposed Moon Treaty." *Houston Journal of International Law,* 2, 3:3.

9. Ibid., 5.

10. Ibid., 7.

11. United States Senate. 1980. "The Moon Treaty." Hearings Before the Subcommittee on Science, Technology, and Space of the Committee on Commerce, Science, and Transportation. 96th Congress, 2nd Session, 15.

12. United Nations Committee on the Peaceful Uses of Outer Space. 1970. U.N. Doc. A/AC.105/C.2/L.71; Cocca, Aldo Armando. 1970. "Legal Status of the Natural Resources of the Moon and Other Celestial Bodies." *Proceedings, 13th Colloquium on the Law of Outer Space.* American Institute of Aeronautics and Astronautics, 146–150; and Menter, Martin. 1980. "Commercial Space Activities Under the Moon Treaty." *Proceedings, 23rd Colloquium on the Law of Outer Space.* American Institute of Aeronautics and Astronautics. 80-SL-14, 35–47.

13. Haanappel, P. P. C. 1980. "Article XI of the Moon Treaty." *Proceedings, 23rd Colloquium on the Law of Outer Space.* American Institute of Aeronautics and Astronautics, 80-SL-10, 29–33.

14. United Nations Committee on the Peaceful Uses of Outer Space. 1972. U.N. Doc. A/AC.105/C.2(XI)WP12.

15. Menter, "Commercial Space Activities."

16. Rusconi, F. G. 1969. "Regime of the Property of the Natural Resources of the Moon and Other Celestial Bodies." *Proceedings, 12th Colloquium on the Law of Outer Space*. American Institute of Aeronautics and Astronautics, 186.

17. Dula, "Free Enterprise," 3.

18. Ibid., 8.

19. Gangale, Thomas. 2006. "Who Owns the Geostationary Orbit?" *Annals of Air and Space Law*. 31:425–446. Montreal, Quebec.

20. Finch, Edward R., Jr., and Amanda Lee Moore. 1980. "The 1979 Moon Treaty Encourages Space Development." *Proceedings, 23rd Colloquium on the Law of Outer Space*. American Institute of Aeronautics and Astronautics, 80-SL-06, 13–18.

21. United States Senate, "Moon Treaty," 63.

22. Finch and Moore, "1979 Moon Treaty."

23. Galloway, Eilene. 1980. "Issues in Implementing the Agreement Governing the Activities of States on the Moon and Other Celestial Bodies." *Proceedings, 23rd Colloquium on the Law of Outer Space*. American Institute of Aeronautics and Astronautics, 80-SL-08, 19–24.

24. Bueckling, Adrian. 1979 (Spring). "The Strategy of Semantics and the 'Mankind Provisions' of the Space Treaty." *Journal of Space Law*, 7:21. Quoted in Finch and Moore, "1979 Moon Treaty," 13–18.

25. United States Senate. "Moon Treaty."

26. Okolie, Charles Chukwuma. 1980. "Legal Interpretation of the 1979 United Nations Treaty Concerning the Activities of Sovereign States of the Moon and Other Celestial Bodies Within the Meaning of the Concept of the Common Heritage of Mankind." *Proceedings, 23rd Colloquium on the Law of Outer Space*. American Institute of Aeronautics and Astronautics, 80-SL-58, 61–67; quoting Almond, Harry H. 1980 (January). "Note." *The International Practitioner's Notebook*, No. 9.

27. United States Senate, "Moon Treaty," 175–178.

28. Michaud, Michael A. G. 1986. *Reaching for the High Frontier: The American Pro-Space Movement, 1972–1984*. New York: Praeger, 90–91.

29. Henson, H. Keith. 1980 (January). "Bulletin from the Moon Treaty Front." *L-5 News*. Quoted in Michaud, *Reaching for the High Frontier*, 91.

30. United States Senate, "Moon Treaty," 233–234.

31. United States Senate, "Moon Treaty," 83.

32. Michaud, *Reaching for the High Frontier*, 91–92; and Godwin, Richard. 2005 (November 16). "The History of the National Space Society." *Ad Astra Online*. Available from http://www.space.com/adastra/adastra_nss_history_051116.html; accessed November 21, 2005.

33. United States Senate, "Moon Treaty," 113.

34. Michaud, *Reaching for the High Frontier*, 92–93.

35. United States Senate, "Moon Treaty," 135.

36. United States Senate, "Moon Treaty," 69.

37. United States Senate, "Moon Treaty," 82–85.

38. United States Senate, "Moon Treaty," 85–103.

39. Broad, William J. 1994 (March 29). "Plan to Carve Up Ocean Floor Riches Nears Fruition." *New York Times.* Available from http://query.nytimes.com/gst/fullpage.html?res=9E00EFDF173FF93AA15750C0A962958260&sec=&spon=&pagewanted=2; accessed February 20, 2008.

40. United Nations. 1994. "Agreement relating to the Implementation of Part XI of the United Nations Convention on the Law of the Sea of 10 December 1982." Available from http://sedac.ciesin.org/entri/texts/acrc/PtXI94.txt.html; accessed October 20, 2008.

41. Galloway, "Issues in Implementing the Agreement."

42. Danilenko, Gennady M. 1989. "Outer Space and the Multilateral Treaty-Making Process." *Berkeley Technology Law Journal,* 4, 2. Available from http://www.law.berkeley.edu/journals/btlj/articles/vol4/Danilenko/HTML/text.html; accessed March 19, 2005.

43. United States Senate, "Moon Treaty," 34.

44. Rosenfield, S. B. 1980. "A Moon Treaty? Yes. But Why Now?" *Proceedings, 23rd Colloquium on the Law of Outer Space.* American Institute of Aeronautics and Astronautics, 80-SL-18, 69–72.

45. United Nations. 1969. "Vienna Convention on the Law of Treaties." 1155 U.N.T.S. 331. Available from http://www.amanjordan.org/english/un&re/un2.htm; accessed July 1, 2005.

46. United Nations. 1963. "Treaty Banning Nuclear Weapon Tests in the Atmosphere, in Outer Space and Under Water." Available from http://sedac.ciesin.org/entri/texts/acrc/Nuke63.txt.html; accessed October 21, 2008.

47. United States Senate, "Moon Treaty," 51.

48. Danilenko, "Outer Space and Multilateral Treaty-Making."

49. Ibid.

50. United Nations Committee on the Peaceful Uses of Outer Space. 1975. U.N. Doc. A/AC.105/C.2 SR.226–245.

51. United States Senate, "Moon Treaty," 40.

5

Moon Myths: What The Moon Agreement Is Not

THE CHALLENGE

Except for the deorbiting of a few spacecraft, nothing has happened on the Moon since 1979, so there has been no hurry to ratify the Moon Agreement. Perhaps the agreement is not dead, but merely sleeping. With the recently developed interest of the United States in returning to the Moon, the time may be approaching when the agreement should be reawakened.

Testifying before the Senate Space Subcommittee on November 6, 2003, former *Apollo 17* Lunar Module Pilot and former Senator Harrison H. Schmitt states:

> On the question of international law relative to outer space, specifically the Outer Space Treaty of 1967, that law is permissive relative to properly licensed and regulated commercial endeavors. Under the 1967 Treaty, lunar resources can be extracted and owned, but national sovereignty cannot be asserted over the mining area. If the Moon Agreement of 1979, however, is ever submitted to the Senate for ratification, it should be deep sixed. The uncertainty that this Agreement would create in terms of international management regimes would make it impossible to raise private capital for a return to the Moon for helium-3 and would seriously hamper if not prevent a successful initiative by the United States Government.[1]

Like many people who have been around the space community for several decades, my impression of the Moon Agreement was negative because everything I had ever read about the agreement was negative. I was a member of the L-5 Society in the late 1970s and early 1980s. Some of my dues funded Leigh Ratiner's lobbying campaign to defeat the Moon Agreement. Then, 25 years later, I did something crazy: I read the actual agreement. I also read its negotiating history and the commentators on both sides. As Martin Menter, vice-president of the International Institute of Space Law (IISL), put it:

> A reading of the [Moon Agreement], without consideration of its negotiated history, reflects some key words or phrases that are without definition, infer a meaning other than intended, ambiguous or not clear in intent. In the COPUOS and its two subcommittees, agreement on a matter under consideration is obtained by consensus; that is agreement is not obtained until no further objection is made. As objections are made, piecemeal changes are suggested. While the intent of a change would be clear at the time it was made, a reader of the entire provision not having the benefit of the detailed consideration accorded the total effort may arrive at a conclusion not in accord with the intent of the provision.[2]

In other words, there is far more to the Moon Agreement than meets the eye. Just as in reading the Constitution of the United States one must at least be cognizant of the settled law that has provided the generally agreed interpretation of that document, one must read the Moon Agreement with a knowledge of other documents that illuminate its language. Article 32 of the Vienna Convention on the Law of Treaties states:

> Recourse may be had to supplementary means of interpretation, including the preparatory work of the treaty and the circumstances of its conclusion, in order to confirm the meaning resulting from the application of article 31 ["General Rule of Interpretation"], or to define the meaning when the interpretation according to article 31:
> (a) leaves the meaning ambiguous or obscure; or
> (b) leads to a result which is manifestly absurd or unreasonable.[3]

It may be pointed out that the United States signed the Vienna Convention in 1970, but has never ratified it. At the time of the Moon Agreement hearings in the U.S. Senate in July 1980, Senator Adlai E. Stevenson III, chairman of the Subcommittee on Science, Technology, and Space, Senate Committee on Commerce, Science and Transportation, asked Robert B. Owen, State Department legal adviser, how that affected the position he had taken on the interpretation of the Moon Agreement in the hearings. Owen's written response, dated August 13, 1980, was that

> [w]hile the United States has not yet ratified the Vienna Convention on the Law of Treaties, we consistently apply those of its terms which constitute a codification of customary international law. Most provisions of the Vienna Convention, including Articles 31 and 32 on matters of treaty interpretation, are declaratory of customary international law.

As a matter of judicial and executive practice, negotiating history, like the legislative history of a statute, is frequently relied upon in U.S. domestic law and in international law to interpret treaties.[4]

Thus the most detailed analysis of the text of the Moon Agreement, without reference to the record of negotiations and the uncontradicted statements of delegations, misses the mark. University of West Florida political science professor David S. Myers's analysis of the agreement is an outstanding example of this. One of his many erroneous conclusions: That's cold

It may be that because of the apparent meaning attached to the "common heritage of mankind" principle, private industry will have little incentive to take the initiative in outer space ventures. That is, if profits from natural resources mining must immediately be passed on to all states, non-governmental entities will be reluctant to become involved. Moreover, the requirement that an international regime be created when exploitation becomes feasible suggests that any private venture in progress would be subject to additional restraints, thus reducing its freedom to an even greater extent. Logically, the apparent legal barriers to freedom of activities for private industry established by this provision will limit the form of economic system in outer space to one of a socialist variety and ultimately to a single, international monopoly. It is surprising that states with a private enterprise tradition have given their initial consent to this provision.[5]

It surprises Myers only because he has focused solely on the text of the agreement to the exclusion of the historical record of its development, thus he has been led to a result that is manifestly absurd and unreasonable. Another opponent of the Moon Agreement, Stanley B. Rosenfield, New England School of Law, also does a great deal to muddy the waters. In addressing the problem of "radically opposed interpretations," to once again use a phrase from Gennady M. Danilenko, Soviet expert on international law:

In each of these problem areas the question is whether the United States view or some other view should prevail. The question is why cannot Art. XI be drafted so that the rights and obligations of all parties are understood in the same manner by all parties? The objection is to an agreement in which major provisions are subject to different interpretations by different states. The objection is to differences which are evident today, but the resolution is left to future speculation.[6]

An examination of the record of the negotiation history will show that COPUOS laid to rest major differences in interpretation and achieved consensus before submitting the agreement to the General Assembly. Most remaining controversy over interpretation arises after the fact from those who have not considered the negotiating history. As to resolution left to future speculation, any adjudication process regarding future disputes would find the agreed interpretation at the time of submittal in the negotiation history and would be obliged to rule on that basis, except where subsequent international agreements had superseded specific provisions of the Moon Agreement in question.

The presentation of Theodore E. Wolcott, a professor at the Hastings College of Law, University of California, is particularly egregious. Barely more than a page of text, it argues no legal points at all. It is a tirade of irrelevancies and errors of fact that would not earn a passing grade in a lower division class. Most of this single page can be boiled down to a stream of condescending unconsciousness:

> The Moon Agreement can be rejected as unrealistic and unsettling. . . .
>
> The final draft has not succeeded on quieting the controversies over the interpretation of its terms and clauses. . . such as "Common heritage of mankind"; "use" and "exploitation of the natural resources of the moon"; "to establish an international regime" to govern the foregoing and assure the "equitable sharing of benefits" with "special consideration" to "the interests and needs of the developing countries as well as the efforts of those countries which have contributed to the exploration of the moon. . ." Keeping pace with a succession of drafts, a spate of learned papers with differing interpretations on the above issue have long poured forth. It may be said with reasonable confidence that the latest version promises grist for the scholarly and other mills for years to come.
>
> It would not be an oversimplification to state that the issues boil down to whether the developed and big countries must wait upon and defer to the developing and small states.
>
> Assuming but not conceding that the time is ripe for some sort of treaty solely dealing with the moon and its environment, it remains questionable whether it would serve any useful purpose to go much further than the principles treaty of 1967. At a minimum Article XI should be omitted as creating nothing but a source of international bickering while at the same time impeding initiative for future development of the moon resources. . . .
>
> Compare the demands of a number of equatorial countries that are claiming sovereignty over and the sole right to that part of outer space above their territories that can be occupied by satellites in geostationary orbit. It is also pertinent to note that whereas the principles treaty provides that outer space should be used only for peaceful purposes, a substantial number of satellites with obvious military implications have been placed in orbit by states party thereto. . . .

But you erroneously believe that there can be but one interpretation based on history. The only thing with power would be the Moon Treaty itself, so interpretation of that is all that matters.

The recommendation by the General Assembly that states ratify the Agreement should be weighed in light of its membership. We are constantly witness to the selective manner in which certain special interest groups, regions and blocs operate in the General-Assembly and in some of the agencies of the U.N.[7]

One may appreciate Wolcott's rhetorical flair, but his paper is a mission without a flight plan; he has flown over a number of targets without inflicting much damage on any. As usual, such invective reveals more about the author than about the subject. Wolcott's antipathy for the Third World and the United Nations is flagrant. His misunderstanding of the role of military space systems is also blatant, and although it is outside the focus of this work, Wolcott's remarks deserve a brief refutation by someone who has worked on some of these military programs. Article 4 of the Outer Space Treaty and Article 3 of the Moon Agreement both declare that the Moon and other celestial bodies shall be used "exclusively for peaceful purposes." Additionally, Article 4 of the Outer Space Treaty provides:

States Parties to the Treaty undertake not to place in orbit around the Earth any objects carrying nuclear weapons or any other kinds of weapons of mass destruction, install such weapons on celestial bodies, or station such weapons in outer space in any other manner.

Therefore, military systems are permitted in Earth orbit so long as they do not carry weapons of mass destruction, and Wolcott's implication that states have routinely violated the Outer Space Treaty is patently false. Nor is a military system necessarily a nonpeaceful use of outer space. For example, the 1963 Partial Nuclear Test Ban Treaty, the 1972 Strategic Arms Limitation Treaty (SALT I), the 1972 Anti-Ballistic Missile (ABM) Treaty, the 1979 Strategic Arms Limitation Treaty (SALT II), all rested on each state's "national means of verification," in other words, surveillance from outer space.[8] Similar language exists in the 1987 Intermediate-range Nuclear Forces (INF) Treaty, the 1991 Strategic Arms Reduction Treaty (START I), and the 1993 Strategic Arms Reduction Treaty (START II), although they also provide for on-site inspection.[9] These military space systems have made nuclear deterrence more stable and less expensive by enabling the rolling back of the arms race and have kept either side from making a fateful strategic miscalculation. These spacecraft have contributed to a more secure and more prosperous world.

There are four principle points of controversy regarding the Moon Agreement:

1. It imposes a moratorium on exploitation of the resources of the moon and other celestial bodies until the establishment of a governing regime.
2. It requires establishment of a governing regime.
3. It prohibits private resource property rights.
4. It allows a governing regime to tax private enterprises.

A common refrain from the Moon Agreement's opponents is to link its language and provisions to those of the 1982 United Nation Convention on the Law of the Sea (UNCLOS III). Richard G. Darman, who in 1980 was at the John F. Kennedy School of Government and 1989 was George H. W. Bush's director of the Office of Management and Budget, issues a fair challenge to the agreement's proponents:

> I end up saying that the probability of the regime to be negotiated if the Moon Treaty were ratified turning out like the emerging seabed treaty is a very high probability; and the burden of argument should be on those who would assert that this is not the case.
>
> It is, however, not a certainty that it would end up like the deepsea mining regime.[10]

This chapter and the following one take up Darman's challenge.

THE COMMON HERITAGE OF MANKIND

At the time of his Congressional testimony against the Moon Agreement, Leigh Ratiner was a partner in the law firm of Dickstein, Shapiro, and Morin. Prior to that, he had been in the U.S. government for 15 years, serving in a number of federal agencies including the Department of the Interior and the Federal Energy Office. From 1969 to 1977 he was a member of the U.S. delegation to the Law of the Sea Conference and for four of those years he was the principal U.S. negotiator for seabed mining issues. As he explains to the Subcommittee on Space Science and Applications, House Committee on Science and Technology:

> It is in part because of this experience at the Law of the Sea Conference that the L-5 Society asked me to appear before you today on their behalf and share with you some thoughts on the recently concluded moon treaty, with particular regard to its relationship to the Law of the Sea Convention which may itself be concluded next year after eleven years of negotiation.
>
> That treaty, Mr. Chairman, gives us the roadmap to the meaning and interpretation which the moon treaty will have, should it ever become the law of the land.[11]

Ratiner's years of experience in dealing with Third World negotiators really shines through in his sensitive retelling of the story of decolonization:

> It has only been since the late 1960's that roughly 90 of the developing countries—there are now about 120 developing countries in the world—became sovereign states, they found that there were economic jeopardies attached to being free. They no longer had the mother countries to provide for their economies. Freedom tasted very good, and all of them opted to remain free and not go back under the wing of the mother country. But they had economic problems of gargantuan proportions.[12]

Thus did the white man unburden himself.

> [a]nd understandably, in the early 1970's, they began to realize that only through concerted political action in the United Nations could they begin to acquire the strength—the voice, if you will—that traditionally had been held by the western industrialized countries and the socialist countries. . . . In 1973, when OPEC taught the whole world a lesson about what collective power can do with respect to the transfer of wealth from the wealthy to the poor, it gave this growing developing country movement in the United Nations tremendous impetus to really begin a rallying cry for the transfer of wealth from industrialized countries to developing countries.
>
> Now, that rallying cry took many different forms in the United Nations organizations where developing countries were represented through this coalition (which came to be called the group of 77, because when the developing countries formed that coalition there were 77). . . .
>
> The Group of 77 in the United Nations developed a manifesto called the New International Economic Order. It is a charter that if I can summarize it—perhaps unfairly, because it's a very long and very complex document—essentially says that those in this world who use the raw materials from those who supply them are going to have to pay for those raw materials dearly.
>
> The New International Economic Order then spread out. Developing country representatives in every different United Nations forum where an opportunity arose adopted that declaration and began to try to proselytize its main features and principles.[13]

The lesson that Ratiner draws from history is clear: if the Senate ratifies the Moon Agreement, the Third World will force future generations of Americans to wait in lines for hours to pay ten dollars per gallon at the gasoline pump.

> Finally, Mr. Chairman, let me say that, in the context of this treaty, there is absolutely no justification for conceding to Third World control the resources of our solar system.[14]

[I]f we sign this treaty and ratify it, we will be sacrificing an interest we cannot even calculate today in terms of the source of the world's resources in the next 100 years. Will they come from outer space, what will those resources be, and what happens to mankind's whole reach for outer space if we essentially put under an international socialist system the development of all the resources in the solar system beyond Earth?[15]

And at the root of this international socialist threat to the American way is the "common heritage of mankind":

Mr. Chairman, the draft moon treaty uses as its legal precedent the 1970 UN Declaration of Principles on the Seabed and Ocean Floor Beyond the Limits of National Jurisdiction which declared the resources of that area to be the "common heritage of all mankind." If we are to fully understand that concept . . . we must first understand the historical context in which it arose and how it fits into a complex international negotiating scenario which is now sweeping through all UN bodies and conferences concerned with economics.

It is therefore incumbent on all of us to scrutinize with the utmost care how the principle that resources are the common heritage of mankind has been interpreted in the Law of the Sea Convention to determine whether the U.S. should now sign a treaty on the moon that contains a significant risk that all of the natural resources of our solar system will be subject to the same international regime as is being contemplated for the bottom of the oceans.[16]

In testimony before the Senate Space Subcommittee, Ratiner suggests that the only way to make sure that the U.S. interpretation of the "common heritage of mankind" prevails for all time is to insist on negotiating a protocol to the Moon Agreement explicitly defining the term before signing and ratifying the agreement. He also makes it clear that in his understanding of global economics, "it would not be possible to do so."[17] Insisting on an impossible protocol is just another way of killing the treaty.

Ratiner does a masterful job. Without a background in space law, he doesn't argue directly against the Moon Agreement. He plays to his strong suit (and to the House and Senate Space Subcommittees' weak suit) by indicting the Convention on the Law of the Sea (on which he had worked for eight years), particularly the development of the meaning of the "common heritage of mankind" in that venue. He implicates the Moon Agreement as guilty by association. Moreover, the Moon Agreement was the nose of the camel under the flap of the tent; if the Senate ratified it, it would find it difficult to reject the Convention on the Law of the Sea (UNCLOS III):[18]

Mr. Chairman, I could perhaps, since I've spent so many years of my life on this subject, go on all day giving you examples of what has happened to the doctrine, common heritage. . . . What we are faced with is a treaty on the Moon that is about to be opened for signature by the United Nations. At least the Law of the Sea Convention is a year or more away.[19]

Representative John Breaux (D-LA) followed his friend Ratiner later in the day, beginning his testimony to the House Space Subcommittee with

You might wonder why someone who's chairman of a fisheries subcommittee has any interest in appearing before your subcommittee.[20]

His purpose certainly was not to discuss fisheries in the Sea of Tranquility, although opponents of the Law of the Sea Convention were so hell-bent on killing the Moon Agreement that one might wonder that they did not assert a direct connection to the lunar maria. In any case, the big money on the street (K Street in Washington, DC) was laid out against the UNCLOS III; the L-5 Society's beef against the Moon Agreement was just a little action on the side.

The technical problem with Ratiner's testimony is that

- It does not interpret the Moon Agreement "in good faith in accordance with the ordinary meaning to be given to the terms of the treaty in their context and in the light of its object and purpose," as required by Article 31, paragraph 1 of the Vienna Convention on the Law of Treaties. The agreement is clear on this point: the term "common heritage of mankind" is to be interpreted in the context of the agreement itself. *But we never ratified that. We don't have to always follow it*
- It does not examine "supplementary means of interpretation, including the preparatory work of the treaty and the circumstances of its conclusion," as required by Article 32 of the Vienna Convention on the Law of Treaties.
- It does not discuss the Moon Agreement in the context of the previous space treaties that it references, which might constitute "supplementary means of interpretation" (the list in Article 32 is inclusive, not exclusive).

Rather, Ratiner's testimony refers to an unfinalized draft treaty with nothing to do with the Moon, the negotiation of which had no connection with the negotiation of the Moon Agreement. Although the Moon Agreement has some superficial similarities to the UNCLOS III—both include the phrase "common heritage of mankind" and both include provisions for international regimes to manage the exploitation of resources in the commons—this does not necessitate that "common heritage of mankind" has the same meaning in both treaties or that the two envisioned regimes will be identical in scope, purpose, structure, or practice. The "relationship to the Law of the Sea Convention" that Ratiner alleges simply does not exist. His assertion that "the draft moon treaty uses as its legal precedent the 1970 UN Declaration of Principles on the Seabed and Ocean Floor Beyond the Limits of National Jurisdiction" is untrue. The Moon Agreement cites its legal precedents in its preamble; the Seabed Declaration is not among them:

> *Recalling* the Treaty on Principles Governing the Activities of States in the Exploration and Use of Outer Space, including the Moon and Other Celestial Bodies, the Agreement on the Rescue of Astronauts, the Return of Astronauts and the Return of Objects Launched into Outer Space, the Convention on International Liability for Damage Caused by Space Objects, and the Convention on Registration of Objects Launched into Outer Space.

Unable to argue against the Moon Agreement using the best evidence (the Article 31 test), the second-best evidence (the Article 32 test), or even the third-best evidence to be found elsewhere in the body of outer space law, Ratiner does the only thing he can: he argues that irrelevant issues are relevant. Most of his testimony hinges on the meaning of the "common heritage of mankind" as developed in the Law of the Sea Conference; however, since the "common heritage of mankind" language in the Moon Agreement finds its expression in the provisions of the agreement itself, whatever it means in the UNCLOS III is irrelevant. Whatever the International Seabed Authority turns out to be is irrelevant, because the Moon Regime, dealing with an entirely different environment, might well evolve in an entirely different direction.* Ratiner presents no legal case at all, but rather a political case. He panders to the fear of "an international *socialist* system" despite the fact that the Union of Soviet *Socialist* Republics objected the most strongly to the inclusion of the "common heritage of mankind" in the agreement. He also panders to the fear of Third World resource embargoes in a United States still reeling from the oil shocks six years earlier, using that experience to build resentment for the newly independent, underdeveloped nations of the world's economic periphery. He does this despite the fact that Argentina (a semi-peripheral industrialized state, independent since 1816) introduced the

---

* Indeed, the UNCLOS III has evolved away from its earlier New International Economic Order planned-economy orientation toward the prevailing neoliberal, laissez-faire, market-economy tenets of the World Bank/International Monetary Fund "Washington Consensus." This is most dramatically illustrated by the 1994 Agreement Relating to the Implementation of Part XI, which significantly rewrote the seabed mining rules.

"common heritage of mankind" term to space law, with the support of the United States (a core industrialized state, independent since 1776) and France (another core industrialized state, independent since 486). Also note that in light of the 1994 Agreement Relating to the Implementation of Part XI of the United Nations Convention on the Law of the Sea, which changes the character of the International Seabed Authority more to the liking of the United States (although it still has not signed the UNCLOS III), even the political arguments that Ratiner raises directly against the UNCLOS III and indirectly against the Moon Agreement are certainly not valid now.

Nor were they were valid then. Ratiner insists on either negotiating a protocol to the agreement, or renegotiating the agreement outright:

> The solution . . . is to negotiate an agreed definition of "common heritage" before accepting it as a binding legal principle.[21]

However, the term was never accepted as a "binding legal principle" in the Moon Agreement, thus Ratiner's insistence on an agreed definition, outside of the language of the agreement itself, is moot. What possible purpose does such insistence serve, except to obfuscate the fact that the term has no independent meaning and is not a "binding legal principle?"* Ratiner repeatedly dismisses provisions of the Moon Agreement against which he can field no argument by calling them "merely hortatory" or indicating that their "advantage is outweighed by the costs of other provisions," yet it is obvious that his primary objection to the agreement—the common heritage clause—is the most hortatory provision of all.[22]

Darman's written statement to the Senate Space Subcommittee discusses the various legal and political meanings of the "common heritage" phrase.

> As a theoretical matter, the phrase might mean anything. If one were to take the plain English language connotation, one might suggest that all of human civilization could be construed as in some sense the common heritage of mankind—although it is not at all clear that one should then leap to the conclusion that all of human civilization must therefore be regulated by a negotiated international regime. . . .

---

* The debate over the meaning of "common heritage" in space law continues to evolve. Harminderpal Singh Rana accommodates a more market-oriented interpretation, whereas V. S. Mani adheres to the centrally-planned NIEO vision of space development.[23] The relevance of this debate to the Moon Agreement is suspect, however, since the agreement requires that the term "finds its expression in the provisions of this Agreement," and in that context, has been characterized as hortatory by many commentators.

If one were to look for legal clarification, one would find that—at least for the time being—the phrase has no formal, widely accepted legal meaning.[24]

The American Bar Association's Section of International Law reports:

There is no generally accepted definition of this term; furthermore, this or any other term may be specifically defined for the purposes of a particular text in which it is used. Although Article 11 does not say expressly what "common heritage" means, it does clearly state that the meaning is to be drawn from the provisions of the Agreement. Particularly as there are numerous relevant provisions within the Agreement to which reference can be made, this explicit direction to derive the definition only from within this text would seem to be legally sufficient to counter any assertion that the draft Law of the Sea Convention must be used as a precedent for the development of the future lunar resources regime. . . .

The essence of the concept of common heritage in the context of this Agreement lies in the existence of common, that is, equally shared, rights to explore and use the moon and its natural resources. It does not, however, connote specific implementing criteria or procedures.[25]

Following Robert B. Owen's testimony, the Senate Space Subcommittee sent him written questions.

Question 1: Is the phrase the "common heritage on mankind" as used in the Moon Treaty in any way related to the use of that phrase in the negotiating text of the Law of the Sea Convention?

Answer: The phrase "common heritage on mankind" finds its meaning only in each particular context in which it is used. Its use in the Moon Treaty is related to the same phrase in the Law of the Sea context in a very general sense; the two negotiations have been largely contemporaneous and, in both, the phrase deal with depletable resources outside the limits of national sovereignty, to be exploited for the benefit of all. However, the detailed meaning and implications of the phrase "common heritage on mankind," as used in the Moon Treaty, are not legally established or determined by the meaning and implementation of that phrase as applied in the negotiating text of the Law of the Sea Convention. Article 11, paragraph 1, of the Moon Treaty expressly denies such a relationship, stating that the phrase as applied to the moon and its natural resources "finds its meaning in the provisions of the agreement."[26]

Just for curiosity's sake, one might ask, how is the phrase "common heritage of mankind" defined in the Law of the Sea Convention? The Senate asked this very question of Marne A. Dubs, chairman of the Committee on Undersea Mineral Resources of the American Mining Congress,

[handwritten margin note: So Moon Agreement is not a Draconian International Law]

and vice-president of the mining firm, Kennecott Development Corporation, who replied,

> The phrase "common heritage of mankind" is not defined in the draft Law of the Sea Convention.[27]

If this were a comedy, the person on-camera would express surprise at this point by spraying his beverage in uncontrollable laughter. Ignoring for the sake of argument that the Moon Agreement mandates that the "common heritage of mankind" must find its meaning in the agreement itself, if the phrase is not defined in the Law of the Sea Convention, that convention has no meaning to convey to the Moon Agreement!

Another interesting exchange in Owen's question and answer follow-up with the Senate Space Subcommittee regards the stance of the only other state to have retrieved natural resources from the Moon (see Table 5.1):

> *Question 4:* What is the Soviet Union's interpretation of the phrase "common heritage on mankind?"

> *Answer:* Early in the Moon Treaty negotiations, the Soviet Union had taken the position that the "common heritage on mankind" was a philosophical, not a legal concept. In its 1973 working paper, the Soviet Union states that celestial bodies are available for the undivided and common use of all States, but are not jointly owned by them. The statements made by their representatives in 1976 and 1977 indicated that they considered the phrase "common heritage on mankind" to be juridically and politically vague, and its inclusion in the treaty "solves no problems." The Soviet Union in 1976 rejected the notion "that space activities should be internationalized and a supra-State nature should be given to whatever body guides those activities." The Soviet representative said that compromise on this matter should be sought through "very accurate interpretation of the concepts used in the draft treaty, on the basis of due respect for the sovereign rights of States participating in space activities." The compromise which was accepted by the Soviet Union in 1979 specifically restricted the meaning of the phrase "common heritage on mankind," as used in Article 11, to the terms of the Moon Treaty itself. The Soviets apparently consider it a political concept devoid of specific legal content and expect the phrase to take on any further content through negotiation.[28]

It is obvious that all of this hand-wringing over the "common heritage of mankind" is not only without any legal foundation, it is also preposterous as a political argument. No opponent of the agreement ever mentions the Soviets' adamant objection to the term.

Table 5.1: Natural Resources Retrieved from the Moon

Mission	Earth Launch	Lunar Landing	Lunar Launch	Earth Landing	Returned Material (kg)
Apollo 11	July 16, 1969	July 20, 1969	July 21, 1969	July 24, 1969	21.55
Apollo 12	November 14, 1969	November 18, 1969	November 19, 1969	November 24, 1969	34.35
Luna 16	September 12, 1970	September 20, 1970	September 21, 1970	September 24, 1970	0.10
Apollo 14	January 31, 1971	February 5, 1971	February 6, 1971	February 9, 1971	42.28
Apollo 15	July 26, 1971	July 30, 1971	August 2, 1971	August 7, 1971	76.80
Luna 20	February 14, 1972	February 21, 1972	February 22, 1972	February 25, 1972	0.06
Apollo 16	April 16, 1972	April 21, 1972	April 24, 1972	April 27, 1972	95.71
Apollo 17	December 7, 1972	December 11, 1972	December 14, 1972	December 19, 1972	110.52
Luna 24	August 9, 1976	August 18, 1976	August 19, 1976	August 22, 1976	0.17
U.S. Subtotal					381.21
USSR Subtotal					0.33
Total					381.54

Source: en.wikipedia.org/

ESTABLISHMENT OF A GOVERNING REGIME

It is interesting to note that although some critics of the Moon Agreement argue the lack of necessity for an agreement governing activities that are not likely to occur for decades, others argue against agreeing to the future establishment of an international regime "as such exploitation is about to become feasible." Some even argue both points: it is too soon to agree on anything now, and we shouldn't agree now to the principle of future negotiation on an agreement. What is wrong with agreeing that one might talk about something in the future?

The truth is that there is no "requirement that an international regime be created," as Myers asserts. Rosenfield makes the same error, and one of Wolcott's few references is Rosenfield:

> What type of authority is to be set up? Such will not be known until negotiation under these provisions is complete. A state party is bound to an international regime, without knowledge of when such will be established, or the contents of such regime.[29]

Ratiner is also in error in his testimony before the House Space Subcommittee:

> Article XI, paragraph 5, of the moon treaty commits the parties "to establish an international regime"[30]

This is simply incorrect, as a reading of the text of Article 11, paragraph 5, immediately reveals:

> States Parties to this Agreement hereby undertake to establish an international regime, including appropriate procedures, to govern the exploitation of the natural resources of the moon as such exploitation is about to become feasible. This provision shall be implemented in accordance with article 18 of this Agreement.

Just means a set of rules

It is in his testimony before Senate Space Subcommittee that Ratiner outdoes himself describing in amazing detail "the Assembly of the International Space Authority" ("the plenary body"), "the Council of the International Space Authority" ("the executive body"), "its Tribunal," and their voting procedures.[31] The entities to which Ratiner refers appear in no document; rather, they are the products of his imagination. This is not even good science fiction; it is space opera.

Regarding such misperceptions, Ernst Fasan of the IISL states:

> We will have to make clear that states *undertake* to establish such a regime and that this means that *not* such a regime would have been established by our agreement. To "undertake to establish" may mean that such a task is

performed successfully by states in a certain space of time, <u>but it also may mean that such an undertaking fails.</u>[32] *But it is the goal, ant cannot guarantee it will fail*

It also means that the U.S. government need not negotiate under the gun, and can hold out for the best deal it can make to protect U.S. private interests. International legal scholar Charles Chukwuma Okolie envisions one power of the Moon Regime to be the execution of a mining rights contract with a corporation after the discovery of a resource had been reported, such contract specifying a zone of exclusive operation for the company. On Earth, such an arrangement would be transacted between a government and a mining company. In the absence of such a Moon Regime, there would be no administrative office with which an enterprise could file a mining claim, thus there would be nothing to prevent another enterprise from tapping into the same resource so long as there were no interference or threat to safety. In a working paper submitted in COPUOS on April 17, 1972, the U.S. stressed the need for states to "recognize the importance of concluding agreements" regarding extraterrestrial natural resources for the purposes of

> economic advancement and for the encouragement of investment and efficient development if utilization of the resources of the Moon and other celestial bodies becomes a reality.[33]

The Moon Agreement is such an agreement. What it is not is a carbon copy of the Law of the Sea Convention. This becomes obvious in Darman's written response to the Senate Space Subcommittee's question, "In what ways does the deep sea mining regime emerging from the U.N. Conference on the Law of the Sea (UNCLOS) differ from the free access regime originally preferred, proposed, and expected by the United States?"

> The institutional arrangements for the proposed new International Seabed Authority do not protect U.S. interests to the extent intended or anticipated.[34]

The Moon Agreement contains no such institutional arrangements for the proposed regime. These details are left to future negotiation of a second treaty.

> Production is directly limited and may be limited further indirectly.[35]

The Moon Agreement contains no direct limitations on production. An indirect limitation might be implied by the requirement to "take measures to prevent the disruption of the existing balance of its environment."

The UNCLOS regime involves a system of mandatory technology transfer.[36]

The Moon Agreement contains no such mandate; it merely states that one of the purposes of the regime to be negotiated shall be "the expansion of opportunities in the use of . . . resources.

Access to natural resources is highly limited.[37]

The Moon Agreement contains no limitations on access except for the requirement that states parties "not interfere with the activities of other States Parties."

[UNCLOS establishes] a globally chartered "Enterprise" in competition with state-sponsored entities on highly advantaged terms.[38]

The Moon Agreement contains no mention of such an "Enterprise."

Unable to complain about the presence of such objectionable specifics in the Moon Agreement, its opponents instead complain about the absence of favorable specifics. This begins as a far weaker argument than a direct criticism of the Moon Agreement, and ends with the logic of this position collapsing on itself. In his arguments against the Moon Agreement in general and the regime in particular, Ratiner insists on vastly greater specificity for such a regime:

States would have absolute and restricted rights to explore for, recover and use space resources;

Each state would determine whether its own nationals could engage in resource activities but would have the responsibility of controlling national activities to ensure compliance with safety, environmental and due diligence requirements of the regime;

States would acquire title to specific resource bodies upon discovery and notification to an international registration office;

Other states would only be bound to respect such resource claims, however, where they satisfied agreed international criteria relating to size, duration and noninterference with other uses;

Disputes over any State's compliance with the regime would be submitted to binding international arbitration;

The States Parties to the regime would engage in periodic consultations concerning scientific and technological developments with a view towards developing recommended standards and practices for avoiding environmental harm and protecting human life and health in carrying out resource activities in space; and

Decision-making procedures would ensure that individual States exercised influence commensurate with their economic interests.[39]

Ratiner insists on specifics that the Moon Agreement never intended to address. The agreement specifically intended for these details to be left for the future negotiation of a Moon Regime when, in NASA Administrator Robert A. Frosch's words, everyone "understands what they are talking about." The essence of Ratiner's case is that the future Moon Regime is a bogeyman in the dark, and we should all be afraid of it unless we can make absolutely sure right now that the bump in the night is really just our puppy dog. It is ironic that the space enthusiasts who aspire to boldly go where no one has gone before fear the outcome of a future negotiation in Vienna. But then, Ratiner's words are not inspired by the pioneering spirit of the High Frontier, but by the acquisitive instincts of the corporate world. For all the vague talk of "close parallels" by opponents of the Moon Agreement between it and the Law of the Sea Convention, the closest parallel of all is between Ratiner's shopping list and that of Dubs:

> Guarantee States and their companies unrestricted rights to explore for, exploit, and use space resources;
>   Allow each State to establish and enforce the basic terms and conditions governing such activities;
>   Create mechanisms for international decisionmaking which give these countries with an economic stake influence commensurate with their interests. . . .
>   Finally, a satisfactory treaty would have to establish agreed and effective procedures for the settlement of disputes.[40]

The two menus are similar because they are drawn up to feed the same corporate appetite. This detailed list of specifications for the regime in effect demands that the regime be precisely defined—in a word, "established"— not as an outcome of a second Moon Agreement, but of the first Moon Agreement. Yet, Dubs believes that "this agreement is premature."[41] The logic of this position is that additional premature specificity makes the agreement less premature.

In addition, although some opponents allude to the Moon Agreement as a Soviet plot, the Soviet Union expressed the greatest opposition to the immediate establishment of a regime, but agreed that the treaty might call for the future establishment of such a regime.[42] The official U.S. position was that an international institutional arrangement would at some point be essential to incentivizing commercial development. Before the idea of a regime became part of the draft agreement, the United States proposed that a future conference would be convened:

> with a view to negotiating arrangements for any international sharing of the benefits of such utilization of the resources of the celestial bodies, bearing in

mind not only the goals of economic advancement, but also to encourage investment and the efficient development of those resources.[43]

Several years of good-faith negotiations ought to produce the international regime envisioned by the Moon Agreement, together with the arrangements regarding the utilization of the resources that it would be charged with administering. As the Canadian representative to COPUOS stated:

> an international regime providing for generally acceptable institutional arrangements would eventually have to be worked out to govern the exploitation of those resources.[44]

International law and political science professor Carl Q. Christol recounts that the chances for a draft agreement being reported out of the Legal Subcommittee looked bleak in 1978. The breakthrough occurred the following year as a result of two compromises. One was the limitation of the meaning of the "common heritage of mankind" to the provisions of the agreement; the other was a consensus to defer the establishment of the Moon Regime to a later, undefined date. Thus the major space powers felt that their interests in exploiting the Moon in the pre-regime period were adequately protected. However, Christol deems Article 11, paragraph 5, to be the heart of the agreement:

> Without the prescription for a future international regime it would not have been possible to achieve the main or other purposes of the Moon Treaty. Without the further provision for the future identification of appropriate procedures there would have been no assurance that machinery would be available to effect the practical dispositions contemplated by the regime's purposes. The agreement that there should be both an international legal regime and an operative instrumentality to advance the specified goals was both logical and practical. . . . [W]ithout Article 11, par. 5, the attempt to bring an orderly legal structure to this aspect of the international law of the space environment would have been incomplete. Without the provision for the regime and the appropriate procedures it would have been possible to speculate that there might not be pursuant to the provisions of Article 11, par. 7, an "orderly and safe development of the natural resources of the Moon," that there might not be a "rational management of those resources," that there might not be an "expansion of opportunities in the use of those resources," that there might not be the creation of a formula for "an equitable sharing by all States Parties in the benefits derived from those resources," and that there might not be an open and orderly process to convert the resources of the Moon and celestial bodies to the service of the values, interests, wants, and needs of mankind.[45]

Question 12 of Owen's follow-up session with the Senate Space Sub-committee and Owen's answer are:

> *Question 12:* How can the Senate be clear and definitive on the point that another agreement dealing with an international regime must come before the Senate for separate consideration and that if the Senate should give advice and consent to the ratification of the Celestial Bodies Agreement, this does not permit the Executive Branch at some future time to conclude an executive agreement concerning an international regime?
>
> *Answer:* The Executive Branch has made clear to the United Nations and the Senate that the agreement establishing an international regime would be submitted to the Senate for its advice and consent to United States ratification, jest as we have sought and obtained advice and consent to United States ratification of the space treaties currently in force. The Senate Foreign Relations Committee should clearly and definitively state its view on the matter in a variety of ways. It might put on the record a statement that advice and consent to the Moon Treaty would not constitute authorization for the Executive Branch to adhere to an agreement concerning an international regime.[46] *But no guarantee?*

At issue here is whether a second Moon Agreement would be handled as a treaty, which requires a two-thirds vote of the Senate, or as an executive agreement, which has no such requirement. Were an understanding included in the Senate's instrument of ratification to the effect that a second Moon Agreement to establish the Moon Regime would require the advice and consent of the Senate, it would remove any legal obligation for the United States to ratify such a future agreement and become subject to such a regime. No legal principle can be construed as obligating the Senate to a preordained course of action. Even without such a formal understanding, Ratiner believes that

> [f]rom a practical point of view, I doubt that the second lunar resource agreement envisioned under Article XI of the Moon Treaty would fail to be submitted to the Senate for its advice and consent. To develop a resource regime, such an agreement would inevitably impose important international obligations upon the United States. In light of the probable substance of these obligations, any Administration would be very unlikely to attempt to implement the agreement without Senate advice and consent, whether or not it could be classified as a "treaty" for the purpose of United States Constitutional requirements.[47]

Since Ratiner allows that the Senate could reject a second Moon Agreement, he demolishes his argument that the United States would be obligated to establish the Moon Regime. Still, he attempts to frighten the Senate with the bogeyman in the dark:

> If we ratify this moon treaty, then we will be obligated under Article XI, paragraph 5, to "establish" a resource exploitation regime. If we then reject the subsequent regime the majority of States Parties negotiate, the political pressure on the U.S. to go along will be tremendous.[48]

Indeed? In view of the many international treaties that have gained wide acceptance and to which the United States is not party, it hardly seems as though the home of the brave allows itself to be pushed around by the rest of the world and caves in to political pressure to go along.

Now, to return to one of Rosenfield's questions, "What type of authority is to be set up?" In his response to a question from the Senate Space Subcommittee, U.S. representative S. Neil Hosenball writes:

> [COPUOS] did not set forth in the treaty nor in the negotiating history what form such a regime should take. What it is to be, its form, its procedures are to be decided at some future time. In French law, it means a system of rules and regulations, but it may also be more generally defined so as to include organizational arrangements.[49]

As an analogy, the General Agreement on Tariffs and Trade (GATT) established a system of rules to regulate international trade, and in that sense it was a regime. However, the International Trade Organization envisioned at the Bretton Woods conference in 1944 did not materialize at the time. The GATT evolved without a standing organization through several rounds of negotiations from 1944 until 1995, when the World Trade Organization was created by the Uruguay Round of GATT negotiations.[50] Given that the global trade system operated for half a century without an administrative organization, it certainly cannot be presumed that the Moon Regime would necessarily take the form of a formal organization at the outset, rather it might evolve into one in the course of many decades as administrative needs evolve. In any case, trepidation over the shape of things to come is not an appropriate basis for opposing the agreement. Ronald F. Stowe, chairman of the Aerospace Law Committee, Section of International Law, American Bar Association, testifying before the Senate Space Subcommittee regarding the Moon Agreement and its purposes, states:

> In the event we are unable to obtain an acceptable outcome from the resource regime negotiations, the United States must be determined enough and independent enough to refuse to become a party to that regime. But we

can make that decision later. There is no advantage or justification for making it now. It is not the present treaty which threatens our interests, but the worst case potential of a future one.[51]

Or as President Kennedy phrased it in his inaugural speech:

Let us never negotiate out of fear. But let us never fear to negotiate.[52]

A MINING MORATORIUM

Danilenko aptly refers to "radically opposed interpretations" of the Moon Agreement. In particular,

[w]hile the Outer Space Treaty proclaims freedom in the use of outer space, which, as generally recognized, includes the freedom to exploit its resources, the Moon Treaty is regarded by many as imposing a moratorium on exploitation of the resources of the moon and other celestial bodies.[53]

Arthur M. Dula uses a semantic analysis of the word "undertake" to construe such a moratorium:

It is arguable that "undertake in Article XI, Paragraph 5, should be read in its obligatory sense. This would place an obligation on the signatory states to establish an international regime that would be equal to their clear obligation to abide by the regime's main purposes. As a practical matter, states engaging in activities affecting the natural resources will determine whether or not their activities are compatible with the major purposes set forth in Article XI, Paragraph 7 either by participating in an existing international regime or by making such determinations unilaterally. In the former case, the international regime must be established before any use may be made of space natural resources other than for the scientific uses specifically permitted by Article VI. In the latter case each state can do whatever it wishes with space resources, which clearly contradicts the purpose of the treaty. Thus, in order to give meaningful effect to the obligations of Article XI, Paragraph 8, the Moon Treaty as a whole contemplates the creation of an international regime prior to allowing the use of space natural resources for the other than scientific and "pilot plant" purposes.[54]

However, as discussed earlier, the word "undertake" does not have the obligatory sense that Dula imputes. The United States stated in COPUOS "that the agreement establishing an international regime would be submitted to the Senate for its advice and consent."[55] This is part of the agreement's negotiation record. No obligation to consent can be presumed;

otherwise the "consenting" party is not a free entity, and consent is superfluous. Given that Dula's construction of "undertake" is incorrect, his construction of a moratorium is baseless.

Other discussions of a supposed moratorium surround the meaning of
the words "exploitation" and "feasible." P. P. C. Haanappel reasonably
construes "exploitation" to mean profit as the principal purpose of the
activity, as distinct from experimental exploitation, and asks

> When is exploitation about to become feasible?
>
> It seems that exploitation about to become feasible, when exploration
> and experimental exploitation have proven that commercial exploitation is
> technically and economically possible. It is submitted that the technical and
> economic aspects must go hand in hand. If exploitation becomes technically
> possible, but its costs would be prohibitive, it seems that further research,
> exploration and technological developments are required to arrive at the
> stage of true feasibility. . . .
>
> Thus, once there is feasibility, States undertake to establish an interna-
> tional regime. . . .
>
> The question then remains, whether full-scale or commercial exploitation
> of the natural resources of the moon is permissible pending the establish-
> ment of an international regime or in the event that States cannot agree on
> such a regime. In other words, does paragraph 5 contain a moratorium on
> exploitation?[56]

A question that Haanappel does not address is how one knows that
commercial exploitation is feasible except by the fact of commercially successful exploitation. A moratorium is logically incompatible with the
requirement to undertake the establishment of a governing regime "as
such exploitation is about to become feasible." Furthermore, what enterprise would engage in such a venture knowing that once it demonstrated
the profitability of its operation it would be required to suspend operations and await the pleasure of the international lawyers and diplomats to
establish a regime before it could resume? Since a moratorium would preclude the establishment of feasibility and therefore the establishment of a
regime, a moratorium cannot possibly be implied by the language of the
agreement. Such an interpretation is manifestly absurd and unreasonable.
Yet Ratiner uses this *ad absurdum* argument, not to falsify a null hypothesis, but, against all logic, to verify it.

> In our judgment, the Moon Treaty will be a potentially crippling disincen-
> tive to investment in these R. & D. activities. Corporations will see it in their
> investment [sic] to await the conclusion of the second lunar resource nego-
> tiation before moving forward to design and test resource exploitation con-
> cepts. Unfortunately, by the terms of the Moon Treaty, these subsequent

negotiations will not commence until exploitation is "about to become feasible"—in other words, until private enterprises or governments make the very financial commitments to R. & D. programs which are deterred by the uncertainties inherent in the provision for a future agreement an exacerbated by the adoption of the common heritage and related principles.[57]

Since the Moon Agreement clearly anticipates the eventual feasibility of exploiting lunar resources, it cannot have the intent of foiling that eventuality. If it cannot have that intent, any interpretation that leads to that effect is incorrect. Any other conclusion is absurd on its face.

Article 6, paragraph 2, of the agreement sets forth the clear right of the extraction of natural resources for specific purposes, thereby codifying what had been established as customary law by U.S control of Apollo mission samples and Soviet control of Luna mission samples.

> In carrying out scientific investigations and in furtherance of the provisions of this Agreement, the States Parties shall have the right to collect on and remove from the moon samples of its mineral and other substances. Such samples shall remain at the disposal of those States Parties which caused them to be collected and may be used by them for scientific purposes. States Parties shall have regard to the desirability of making a portion of such samples available to other interested States Parties and the international scientific community for scientific investigation. States Parties may in the course of scientific investigations also use mineral and other substances of the moon in quantities appropriate for the support of their missions.

If exploitation means that the principal purpose of the activity is profit, commercial exploration counts as exploration and not exploitation, and is therefore permitted. The purpose of such missions would be to explore the feasibility of exploitation, not exploitation per se, thus they would be permitted to "use mineral and other substances of the moon in quantities appropriate for the support of their missions." This is a form of "scientific investigation" in that it is science applied to the purpose of making a profit some time in the future. Applied science is science. It must also be considered that this paragraph not only refers to "carrying out scientific investigations," but to the "furtherance of the provisions of this Agreement." It is only logical to construe the latter phrase to mean activities that further the objective of advancing technology to the point at which "exploitation is about to become feasible," a state of affairs that key provisions of the agreement are clearly meant to address.

Additionally, as Haanappel observes, "in the event that the treaty drafters would have wished to create a moratorium, they would have done so in more explicit terms." Still, despite the obvious illogic of a moratorium, there would be some room for doubt on this subject if it had not been raised during negotiations. Early in the negotiations, however, statements by the COPUOS representatives of Mexico, India, and Iran put forward the claim that the obligation to establish an international regime amounts to recognition of the moratorium.[58] From beginning to end, the United States opposed any suggestion of a moratorium:

> The treaty relating to the Moon could not reasonably require that exploitation must await the establishment of the treaty-based regime.[59]

Edward R. Finch Jr. and Amanda Lee Moore state:

> Two factors lead to the conclusion that no moratorium on the exploitation of lunar resources was either intended or established. First, neither this Article nor the Treaty establishes the right of States to exploit lunar resources. That right was recognized in Article 1 of the 1967 Outer Space Treaty. Paragraph 5 cannot, therefore, be read as a conditional grant of a new right, and it does not purport to be a limitation of an existing one. Second, if the drafters had intended to adopt a moratorium, they could easily have done so. The issue was expressly discussed before the Legal Subcommittee and the [COPUOS]. However, a review of the negotiating history reveals that language specifically calling for a moratorium was, in at least two instances, rejected during negotiation of the Moon Treaty.[60]

Also, any attempt to construe a moratorium where none is explicitly imposed runs counter to the basic principle of international law: that which is not prohibited is permitted. Nowhere in the Moon Agreement is the exploitation of natural resources prohibited in advance of the establishment of a governing regime, therefore such exploitation is permitted.

Rosenfield, an opponent of the agreement, states:

> Whether Art. XI(5) requires the international regime to be in place before resources can be exploited is still being debated, The United States position is that it does not, as indicated by the statement of the U.S. representative delegate placed in the record. . . . Not all states agree with the U.S. interpretation.[61]

However, Rosenfield fails to document when the statement he cites was placed in the record, which states did not agree, when any statements of disagreement were made, and in what forum these statements were made. This much is fact: the statement of U.S. representative Hosenball on July 16, 1979, once the committee had finalized the text of the agreement, was the last word on the subject in the official record of the negotiations:

[The agreement] places no moratorium upon the exploitation of the natural resources on celestial bodies, pending the establishment of an international regime. This permits orderly attempts to establish that such exploitation is in fact feasible and practicable, by making possible experimental beginnings and, then, pilot operations, a process by which we believe we can learn if it will be practicable and feasible to exploit the mineral resources of such celestial bodies.[62]

As Hosenball testifies before the House Space Subcommittee:

These statements by the United States were not contradicted and constitute a part of the legislative history of the treaty negotiations.[63]

The only time Ratiner comes close to making a legal argument in his House Space Subcommittee testimony is when he claims that in Hosenball's statement:

His own definition of the Moon Treaty excludes the permissibility of exploitation, and there is no question, using the civil law principle of *a contrario sensu* or the American legal principle of negative pregnant, that in this statement alone we have probably said to the rest of the world we do not think exploitation is permitted under the Moon Treaty.[64]

However, Ratiner offers no explanation of this assertion, and since what he says is contrary to a common sense reading of Hosenball's statement, his negative interpretation miscarries. Dula takes another crack at it and comes to an equally erroneous conclusion regarding Hosenball's statement:

The United States' effort to preserve its legal rights to engage in resource development under the treaty clearly stops short of full-scale exploitation. Moreover, there is a strong legal inference, arising from paragraph of Article VI and paragraph 8 of Article XI, that commercially-oriented enterprises are even barred from engaging in the kind of experimental or pilot operations in Mr. Hosenball's statement above. The treaty permits the use of resources in "scientific investigations." To the extent that it excludes research and development activities undertaken by a commercial entity in the hope of future profit, paragraph 2 of Article VI would prohibit such an entity's using resource samples collected from the Moon and other celestial bodies either for research and development or for support of its missions. Paragraph 8 of Article XI reinforces this apparent prohibition on the conduct of interim resource activities by states and persons who are not pursuing scientific purposes.[65]

Apparently, Dula believes that "scientific investigation" and "commercial research" are mutually exclusive terms. From this, we may confidently conclude that neither Nikola Tesla nor Guglielmo Marconi were scientists because both conducted their research in the hope of future profit.

What possible motivation would the United States have for "stopping short" in preserving its legal rights? Dula's interpretation certainly makes no sense from a policy standpoint. And, again, it flies in the face of a fundamental legal principle: that which is not prohibited is permitted. Simply because "commercial exploitation" or "full-scale exploitation" is explicitly permitted is no basis for asserting that they are prohibited. In Dula's case, the source of the error is clear; he simply misrepresents Hosenball's statement by omitting his assertion that the agreement "places no moratorium upon the exploitation of the natural resources." This includes whatever adjective one cares to place before "exploitation," "commercial," "full-scale," et cetera.

From this, Eilene Galloway concludes:

> It is clear that there is no legal moratorium in the Moon Agreement prohibiting the exploitation of natural resources of the Moon and other celestial bodies between the present time and the establishment of an international regime, but some would argue that it would be even more certain if the wording of paragraph 56 had been included in the treaty text.[66]

Returning to Owen's follow-up with the Senate Space Subcommittee:

> *Question 7:* If the United States signed and ratified the agreement, would this give other nations any control over the timing and direction of the U.S. private sector engaging in and expanding the commercial uses of space?

> *Answer:* No. The Agreement establishes no legal moratorium on resources exploitation of other commercial uses of outer space and gives no other nation or group of nations control over the timing and direction of United States space programs. The United States would not be obligated to adhere to a future international regime for resource exploitation if it contained unacceptable provisions in this or any other respect, and nonadherence to such a regime would not in any other way limit the right of the United States to exploit nonterrestrial natural resources.[67]

Hosenball's statement in COPUOS regarding the absence of a moratorium was uncontradicted:

> The draft agreement . . . as part of the compromises made by many delegations, places no moratorium upon the exploitation of the natural resources on celestial bodies, pending the establishment of an international

regime. This permits orderly attempts to establish that such exploitation is in fact feasible and practicable, by making possible experimental beginnings, then, pilot operations, a process by which we can learn if it will be practicable and feasible to exploit the mineral resources of such celestial bodies."[68]

Dula remarks that Hosenball's statement seems "intended to contradict the clear language of the treaty regarding exploitation of space natural resources."[69] Here Dula defends the clarity of a document that elsewhere he dismisses as "vague."[70] More importantly, Dula entirely misses the significance of Hosenball's statement. Since the statement was uncontradicted and appears in the COPUOS final report on the Moon Agreement, it represents the consensus of COPUOS regarding the interpretation of the agreement and, pursuant to Article 32 of the Law of Treaties, this interpretation takes precedence over any other interpretation regarding a moratorium.

Ironically, one of the strongest claims by Moon Agreement opponents is espoused by those who shy from asserting that there is a *de jure* mining moratorium, but at the same time theorize that it creates a *de facto* moratorium. They claim that the ambiguity of the phrase "common heritage of mankind" and of the power and scope of the regime to be born of future negotiation, combined with their possible link to the Law of the Sea Convention create so much uncertainty as to deter private enterprise from making the tremendous capital investments necessary to develop extraterrestrial resources. This is not a strong claim because it is based on anything real, but rather, because it is not. It is like a self-fulfilling prophesy, as when Federal Reserve chairman Alan Greenspan spooks Thomas Friedman's "electronic herd" and causes the Dow Jones Industrial Average to drop hundreds of points simply by asking:

> [H]ow do we know when irrational exuberance has unduly escalated asset values, which then become subject to unexpected and prolonged contractions as they have in Japan over the past decade?[71]

It can hardly be denied that at some level, economies operate as a confidence game; when investors lack confidence, they are hesitant to play the game. Thus, the more that Moon Agreement opponents say that the uncertainties about the agreement create a de facto moratorium, the more likely it is to be true, because they are doing as much as anyone to create these uncertainties in the minds of investors regarding their protections in the Moon Agreement. They are "talking down" the lunar economy before it can even get off the ground.

RESOURCE PROPERTY RIGHTS

Hosenball made the U.S. position on resource property rights clear early in the negotiation of the Moon Agreement:

> [T]he words "in place" . . . are intended to indicate that the prohibition against assertion of property rights would not apply to natural resources once reduced to possession through exploitation either in the pre-regime period or, subject to the rules and procedures that a regime would constitute, following establishment of the regime.[72]

Dula portrays Hosenball's statement as unilaterally contradicting "the clear meaning of the words 'in place'"; however, Dula does not say what he believes that "clear meaning" to be in a document he accuses of being vague, nor why Hosenball's statement contradicts it.[73] Seemingly, the only unilateral interpretation permitted by Dula is his own, whatever that might be. In any case, as Dula himself states, Hosenball's statement "drew no response, and this silence is . . . a part of the history of the treaty." This fact proves that Hosenball's statement was not "unilateral"; rather, it expressed the consensus of the COPUOS Legal Subcommittee. An uncontradicted statement captures the consensus.

H. L. van Traa-Engelman of the Netherlands, who favors the Moon Agreement, nevertheless expresses concerns that it will put a damper on commercial exploitation:

> Following the context of the relevant articles in the Moon Treaty there exists no explicit prohibition to exploit the moon resources before the existence of an international regime. But the expressed impossibility to obtain property rights outside of an international regime, imposed by Article XI.3 together with the acceptation of "the Common Heritage of Mankind" principle of Article XI.1 will definitely inhibit actual exploitation by any nation, party to this treaty or private enterprise in any state, party to this treaty.[74] *Too strong language. Many are not deterred*

But, is there an "expressed impossibility to obtain property rights outside of an international regime?" The legality of resource appropriation (as opposed to real property rights) can be—and has been—derived indirectly from the Outer Space Treaty; however, the Moon Agreement specifies in Article 11, paragraph 3:

> Neither the surface nor the subsurface of the moon, nor any part thereof or natural resources in place, shall become property of any State, international intergovernmental or non-governmental organization, national organization or non-governmental entity or of any natural person.

[handwritten: But they all start M place]

Since the prohibition only applies to resources in place, resources extracted from the Moon may become property. Haanappel comments:

> The words "in place" are of paramount importance. An *a contrario* reasoning shows that natural resources on the moon, once no longer in place but removed, may become the property of States, organizations and natural persons. This may happen both during the pre-regime period pursuant to Article XI(4) and Article VI(2), and during the regime period pursuant to Article XI (5) (7), but during this latter period as qualified by the terms of the international regime.[75]

When Hosenball proposed the insertion of the term "in place" on April 17, 1973, he explained that the purpose was

> to indicate that the prohibition against assertion of property rights would not apply to natural resources once reduced to possession through exploitation either in the preregime period or, subject to the rules and procedures that a regime would constitute, following the establishment of the regime.[76]

[handwritten: OK, I get it now]

The committee accepted the U.S. amendment without objection to Hosenball's statement, thus this interpretation reflects the consensus of the committee, and any contradictory interpretation is incorrect.

Shortly after the Moon Agreement was opened for signature, Soviet law professor R. V. Dekanozov averred:

> As regards the resources of celestial bodies, their juridical nature differs from that of the celestial bodies proper. The juridical nature of the latter is determined by principles of non-appropriation and common use. It is quite different with the resources of celestial bodies. The principle of non-appropriation applies to them, in particular, solely until they have not [sic] been extracted. It is this interpretation that should be given to Art. XI, para. 3 of the 1979 [Moon] Agreement. . . . What is implied here, as testified by the words "natural resources in place" and the general meaning of Art. XI, are natural resources which have not yet been alienated from the lunar territory, lunar surface or subsurface and make up with them one whole. Evidently, those resources which have already been extracted may be considered property.[77]

[handwritten: So you can't just claim a gold field]

So we see that the only two states that had at that point—and have ever—extracted resources from the Moon were in complete agreement regarding the legitimacy of such resources as property.

Furthermore, Article 6, paragraph 2, provides in part:

> In carrying out scientific investigations and in furtherance of the provisions of this Agreement, the States Parties shall have the right to collect on and remove from the moon samples of its mineral and other substances. Such

samples shall remain at the disposal of those States Parties which caused them to be collected and may be used by them for scientific purposes.

Even Moon Agreement opponent Rosenfield admits, albeit with some demurral:

This is the first outer space treaty specifically granting the right to collect and remove and use substances from outer space. While minerals are not specifically mentioned in the 1967 Outer Space Treaty. It must also be noted that the practice to date has established the custom of collection and removal of samples.[78]

Nevertheless, the provision is a clear statement for property rights over extracted resources, which is lacking in the Outer Space Treaty. Haanappel believes that the Moon Agreement is stronger on both explorative and exploitative resource property rights than the Outer Space Treaty is

the right of exploitation is recognized, something that the Outer Space Treaty had not done specifically. Article I(2) of the Outer Space Treaty recognizes the right to exploration and use, but is silent on the question of exploitation giving rise to different interpretations in this respect.

Finally . . . Article VI(2) [clarifies] the status of samples removed from the moon. Under the regime of the Outer Space Treaty and before the drafting of the Moon Treaty the right to collect samples was not universally recognized.[79]

Menter, of IISL, concurs:

It has been the reported position of some States and attorneys that exploitation of the natural resources of the moon would not be lawful under the present state of Space Law. Stated another way, the doubt that existed as to the lawfulness of exploitation of the moon's resources will be removed by the [Moon Agreement's] providing for exploitation. Hopefully, this will encourage private sector investment in such endeavor. The need to encourage such investment has been noted as a must item for subsequent conferences seeking to establish the international regime.[80]

Hosenball's 1979 statement in COPUOS was uncontradicted:

Article XI, paragraph 8, by referring to Article VI, paragraph 2, makes it clear that the right to collect samples of natural resources is not infringed upon and that there is no limit upon the right of States parties to utilize, in the course of scientific investigations, such quantities of those natural resources found on celestial bodies as are appropriate for the support of their missions. We believe that this, in combination with the experimental and pilot programs, will foster and further, and perhaps speed up, the possibility of the commercial or practical exploration of natural resources (COPUOS 1979; USHR 1979, 86).[81]

Regarding Article 11, paragraph 5, and its implications for private enterprise, Okolie adds:

[W]e do not read any meaning into these lines that could be interpreted as barring participation of national and transnational corporations. By calling for a universal conception of the common ownership of outer space by all sovereigns, the treaty seems to give rise to a legal interpretation that exploitation of the natural resources of the Moon should be conducted through the aegies [sic] of national government or through the operation of the law of the place of legal existence of such corporations, namely a state party to this treaty. Thus, the meaning we derive from this paragraph is that transnational corporations have the right to participate in space exploration and exploitation, but must do so through and with the protection of a state party to the treaty.[82]

However, Ratiner warns of the detrimental effect on commercial activity coming from the duty to "inform the Secretary-General of the United Nations as well as the public and the international scientific community, to the greatest extent feasible and practicable, of any natural resources they may discover on the moon:"

Article XI, paragraph 6, of the moon treaty does require States Parties to disclose resources discovered on celestial bodies. This provision is arguably only a slight expansion of the existing obligation under Article XI of the Outer Space Treaty to inform of the results of outer space activities. To the extent the moon treaty makes the disclosure obligation more specific with respect to resources, however, it is probably detrimental to U.S. interests in commercial resource development. The reason is that mandatory disclosure of resource discoveries is a disincentive to private exploration for resources for commercial purposes.[83]

This is yet another of Ratiner's "worst case" readings of the agreement's language. Is it not true that when a private entity files a mining claim it informs a public entity, to the greatest extent feasible and practicable, of any natural resources it may have discovered? If there is no public disclosure, there can be no legal protection. One of the purposes of disclosure "to the greatest extent feasible and practicable" is to characterize the extent of the discovered resources and to delineate the claim thereto. This purpose is inherent in the right to be free from interference under Article 8, paragraph 3. Ratiner, however, disagrees on this point:

I agree that the principle in Article IX, paragraph 1, which limits the size of areas needed for stations could prove a useful foundation for requiring in a later regime that resource claims must be diligently worked.[84]

Ratiner is so intent on picking apart the Moon Agreement that he fails to see the whole picture: Article 11, paragraph 6, provides for public disclosure, which establishes the basis of a claim; Article 9, paragraph 1, limits the size of a claim to that necessary to the mission of the station; and Article 8, paragraph 3, establishes the right to be free from interference while working the claim.

Regarding the protections that the Moon Regime might provide, Hosenball testifies:

[A]rticle 9, paragraph 1, taken in conjunction with article 9, paragraph 2—shall not impede free access to all areas of the Moon—article 8, paragraphs 1, 2, and 3—pursuit of activities anywhere on or below its surface; place facilities, stations and installations on or below the surface of the Moon; and states shall not interfere with activities of other state parties on the Moon—and article 11, paragraphs 2 and 3—no sovereignty, no ownership—if applied to a mining facility in effect establishes procedures which I believe parallel mining operations on public lands in the United States; that is, locating an area having potential mineral resources, staking out the claim, filing the location of the claim with a central office, and the requirement to work the claim or it is considered abandoned.

The filing of the location of a claim does not transfer ownership of the public land to the claimant and similarly under the Moon Treaty the use and occupancy of an area of the Moon does not transfer ownership of that area of the Moon.

The noninterference provision of the treaty does in effect give the state party, or private enterprise party acting under authority of a state party, the equivalent of an exclusive privilege to mine the claim at the location reported to the Secretary General.

These treaty provisions also limit the area that can be mined to that which is required to conduct the mining operation, thus preventing a single state from excluding others from areas it is not using for such operations. Thus the Moon Treaty does contribute to the continuing development of space law and does provide more certainty than currently exists under the 1967 Outer Space Treaty.[85]

Viewed from this perspective, the failure of the Moon Agreement to be widely ratified has left an unsolved problem for prospective planetary enterprises, which their governments will need to address at a future date: the ability of entities to acquire clear title to extracted natural resources.

DEATH AND TAXES

Opponents of the Moon Agreement raise the concern that the Third World will demand such a large piece of the action that no private interests will invest in a lunar mining venture. In other words, the fear of being taxed to death will discourage investment. In his testimony before the Senate Space Subcommittee, Frosch voices skepticism with respect to the argument that general principles such as "common heritage of mankind" and "for the benefit and in the interests of all countries" would have a chilling effect on private investment:

> Only if you insist on being chilled. I am not all that sympathetic with that. It is so general a statement that it is hard to be certain that it is, in fact, something that will be chilling. Subject to future negotiation. Everything is subject to tax somewhere and all taxes in the United States are subject to the future actions of the U.S. Congress. So no investor is sure of what will happen to his investment 5 years from now or 10 years from now. It is simply not clear to me that this general statement that you should be for the general good, is in fact, a more chilling thing than the prospect of future taxes in the United States.
>
> But I presume, if I understand the structure of the treaty, that whatever regime were to be negotiated, would be a treaty matter and that in itself would have to be ratified by the U.S. Senate. That is, it would not simply be a matter of ratifying this treaty and then all else following automatically without reference to Congress. In that sense, it would be a matter of Congressional legislation.[86]

While Ratiner, Rosenfield, Wolcott, and others decry the United States being taken to the cleaners by the Third World in the Moon Agreement, Damodar Wadegaonkar, writing from the Third World, voices the opposite criticism:

> [I]t can be seen that the concessions made by the Space Powers to the demands of the developing countries for establishment of an international regime does not represent more than an appeasement in quest of that elusive concept, namely, consensus. Even the concept of equitable sharing is already circumscribed by the proviso inserted by the Space Powers that special consideration would be given to the Space Powers whilst apportioning the expected benefits. In other words, the Space Powers were reluctant to let go of their pound of flesh.[87]

Is it reasonable to look for the truth somewhere between these two extreme positions?

Before considering the agreement language itself and its implications, two points can be considered: one retrospective, one prospective.

First, in the latter half of the twentieth century, governments of the eastern Asian "tiger economies" of Thailand, Singapore, Taiwan, and South Korea, and in the free-market industrial states of Japan, North America, and western Europe, have often created tax incentives—and in some cases have provided investment capital—to stimulate the development of new technologies for economic development. Indeed, in eastern Asia this strategy has been raised to a fine art, and has led to the rise of the term "developmental state" to describe the formerly underdeveloped Third World states that have graduated into the industrialized First World. Thirty years ago, when the Moon Agreement was being negotiated, the only example of a developmental state was "Japan, Inc.,"[88] which was not a Third World country, but a First World country that had recovered from the devastation of World War II. In view of the intervening three decades of economic history, are the fears that were expressed by opponents of the Moon Agreement (whether or not valid then) valid today? If the New International Economic Order is to legitimately hold to its first adjective "new," it must be updated to take into account the "developmental state" experience. Likewise, the "common heritage of mankind" concept, whatever it might have meant to particular delegations and commentators when the Space Shuttle was still on the drawing boards, is going to have a different meaning in the future.

This leads to the second point: the Third World, which might demand a piece of the action on the Moon and other celestial bodies, has been shrinking. The eastern Asian "tigers" are now First or Second World states, the Chinese economy will surpass the U.S. economy in the near future, and India is a launching state as well as software developer to the world. So, when the exploitation of the Moon "is about to become feasible" and the negotiation of a governing regime can begin, who will still be in the Third World? Of course, this is not to suggest that there do not remain regions of crushing poverty, malnutrition, and lack of education, and that there are not problems of uneven economic development remaining to be addressed. It is only to suggest that if these problems are not substantially reduced when the time arrives to negotiate a governing regime to manage lunar resources, we have not been doing a very good job of managing Earth's resources.

Turning now to the language of the Moon Agreement and its various interpretations, Article 11, paragraph 7(4), states:

> An equitable sharing by all States Parties in the benefits derived from those resources, whereby the interests and needs of the developing countries, as well as the efforts of those countries which have contributed either directly or indirectly to the exploration of the moon, shall be given special consideration.

Dula speculates that

> the principle of "equitable sharing of benefits" could be interpreted to require a system of international taxation of any profits made by commercial resource developers. Since the term "benefits" is not restricted to the financial realm, the principle could dictate mandatory transfer to all countries of the technology used to exploit the resources.[89]

Maybe it could, but it need not. Such speculation does not build a persuasive case. Again, the Moon Agreement contains no language regarding technology transfer; this is an issue that has been speciously transferred from the Law of the Sea Convention, which does contain such provisions. It is noted with interest that Dula cites as the source of this speculative argument Leigh Ratiner's August 15, 1979 letter to L-5 Society president Carolyn Henson.*

Finch and Moore conclude that the Moon Agreement's language concerning "'sharing benefits' and 'equitable sharing' refer to an international ordering of an international resource," as opposed to national appropriation, which is forbidden by the Outer Space Treaty:

> In general, the Moon Treaty envisages few limits on the exploitation of lunar resources beyond the protection of the environment and safety. An international ordering by a future regime is just the sort of climate which should encourage investments.[90]

Haanappel notes that "equitable sharing" does not mean "equal sharing," and that paragraph 7(4) "makes it clear that a balance may be struck in terms of benefits between developing countries on the one hand and countries having participated in the exploration of the Moon."[91] Galloway makes a similar observation:

> Equitable sharing is not the same as equal sharing, and in this context suggests some arrangement such as that in the Convention on the International Maritime Satellite Organization (INMARSAT), which provides, in Article 8, that "The Organization shall be financed by the contributions of Signatories. Each Signatory shall have a financial interest in the Organization in proportion to the investment share which shall be determined in accordance with the Operating Agreement."[92]

---

* Ratiner's letter is cited in seven footnotes to Dula's article. It is odd that a non-scholarly source would be so heavily relied on in a law journal article. Dula also cites the House Space Subcommittee hearings of September 5–6, 1979, by which time the L-5 Society had hired Ratiner to lobby against the Moon Agreement.

Thus, a State Party's "equitable share" in the extraterrestrial mining regime would be proportional to its investment. Menter also arrives at a similar conclusion:

> Justinian is said to have defined "equity" as "to live honestly, to harm nobody and to render every man his due." While the subsequent agreement for the international regime may establish criteria, it is conceivable that the share "due" to a non-contributing state may be determined to be zero.[93]

All of this suggests that just as here on Earth, on the Moon, TANSTAAFL: "There ain't no such thing as a free lunch. . . . One way or other, what you get, you pay for."[94]

A final observation regards the power of the United Nations (or any regime that might be established under its auspices) over states, juridical persons (corporations), and natural persons. States are subjects of international law; generally speaking, juridical persons and natural persons are not. Exceptions occur under certain trade agreements or for international crimes. For example, under the North America Free Trade Agreement (NAFTA), Methanex Corporation, a Canadian manufacturer of methyl tertiary-butyl ether (MTBE) was able to bring a $970 million suit against the state of California when it banned the carcinogen as a fuel additive.[95] The prosecution of former Yugoslav president Slobodan Milosevic is another such example.[96] Thus, were the United States to become party to the Moon Agreement, and were an international regime to be established to govern the exploitation of resources on the Moon and other celestial bodies, such a regime would apply directly only to the U.S. government. The Moon Regime would apply only indirectly to juridical persons and natural persons under the jurisdiction of the United States, in that the United States would bear international responsibility for the activities of such persons. If the United States were to incur a financial obligation under the Moon Regime, it would not be obliged to pass on that obligation to the private interest conducting the actual mining operation on the Moon. Such actions would be at the discretion of the U.S. government. To incentivize lunar development, the U.S. government might well decide to pay whatever was due to the Moon Regime, letting the private U.S. interests off the hook. The U.S. government might also provide tax incentives, loan guarantees, and study grants for lunar development. In short, the United States might well choose to craft a policy as a transplanetary developmental state. All of this is possible under the Moon Agreement.

Dula, a patent lawyer, focuses on the minutia of his field and entirely misses the larger issues:

It may be suggested that investors could simply call whatever activities they decide to undertake in space or on the Moon "research and development," "experimentation," or "pilot plant operations." Unfortunately, this is not advisable because of differences in the legal consequences of experimental and commercial use. United States tax laws and regulations controlling investment in commercial operations differ greatly from the rules that apply to investment in experimental research or pilot plant operations. . . . If an operation consistently returns a profit to its owner, then it is commercial for the purpose of U.S. tax and investment law.

   U.S. patent laws also differentiate between commercial and experimental purposes.[97]

Although Dula may be well versed in the patent law and tax code of 1980, he certainly has no idea how these might be tailored to incentivize commercial development on the Moon in 2020 or at some other future point.

As Hosenball points out, risk is in the eye of the beholder:

I believe it is up to industry, frankly, to decide what risk they are willing to take. There is nothing in this treaty that prohibits them or restricts them from taking that risk. I think they will have to decide at what point in time, based in investment, they would be willing to take the risk.[98]

Space enthusiasts are fond of saying, "The meek shall inherit the Earth; the rest of us will go to the stars." It would seem that this is only the case if the heavens are tax havens.

The Moon Treaty, if ratified, would be U.S. Federal law and could be enforced by the Federal courts. A group . . . could bring suit to obtain an injunction against the U.S. Government, the company involved, or both.[99]

Can Dula point to any provision in the Moon Agreement that is self-executing? If the agreement is non-self-executing, any affect on U.S. private legal persons would only come from whatever Federal statutes were enacted to give effect to the Moon Agreement in municipal law. Absent this, no suit would have legal standing.

   The disincentives to private investment that opponents of the Moon Agreement point to are more perceived than real, and are perceived by those who are predisposed to misperceive. The truth is that it is up to private industry to step up to the plate and take the risks to develop the Moon and other celestial bodies, which it will do when it sees a reasonable probability of an adequate return on investment over an acceptably short period of time. As a credible level of interest in such activities develops, private industry is certainly capable of using—and certainly will use—its lobbying influence to obtain federal legislation shielding it to an acceptable degree from whatever concerns it might have with regard to the Moon Regime. But just as there should be no free lunch for non-participating states, neither should there be a free lunch for private enterprises operating on the Moon and other celestial bodies. Governments have gone to some effort and expense to establish a legal regime in which private enterprises can extract profits, and governments can legitimately expect a return on their investment in the form of taxing those profits. Governments should also expect to levy taxes to defray the continuing cost of administering the regime and providing other public goods that facilitate profitable operations. Anything less would not be "equitable."

As science fiction novelist Robert A. Heinlein's unruly lunar colonists put it, "TANSTAAFL!"* That acronym cuts more ways than one.

NOTES

1. Schmitt, Harrison H. 2003 (November 6). "Summary Testimony." Subcommittee on Science, Technology and Space of the Senate Commerce, Science, and Transportation Committee. Available from http://www.globalsecurity.org/space/library/congress/2003_h/031106-schmitt.doc; accessed December 22, 2005.

2. Menter, Martin. 1980. "Commercial Space Activities Under the Moon Treaty." *Proceedings, 23rd Colloquium on the Law of Outer Space.* American Institute of Aeronautics and Astronautics. 80-SL-14, 35–47.

3. United Nations. 1969. "Vienna Convention on the Law of Treaties." 1155 U.N.T.S. 331. Available from http://www.amanjordan.org/english/un&re/un2.htm; accessed July 1, 2005.

4. United States Senate. 1980. "The Moon Treaty." Hearings Before the Subcommittee on Science, Technology, and Space of the Committee on Commerce, Science, and Transportation. 96th Congress, 2nd Session, 19.

5. Myers, David S. 1980. "The Moon Treaty in Legal and Political Perspective." *Proceedings, 23rd Colloquium on the Law of Outer Space.* American Institute of Aeronautics and Astronautics, 80-SL-14, 49–55.

6. Rosenfeld, S. B. 1980. "A Moon Treaty? Yes. But Why Now?" *Proceedings, 23rd Colloquium on the Law of Outer Space.* American Institute of Aeronautics and Astronautics, 80-SL-18, 69–72.

7. Wolcott, Theodore E. 1980. "Reaching for the Moon." *Proceedings, 23rd Colloquium on the Law of Outer Space.* American Institute of Aeronautics and Astronautics, 80-SL-18, 87–88.

---

* "There ain't no such thing as a free lunch."

8. United Nations. 1963. "Treaty Banning Nuclear Weapon Tests in the Atmosphere, in OuterSpace and Under Water." Available from http://sedac. ciesin.org/entri/texts/acrc/Nuke63.txt.html; accessed October 21, 2008; United Nations. 1972. "Interim Agreement Between the United States of America and the Union of Soviet Socialist Republics on Certain Measures With Respect to the Limitation of Strategic Offensive Arms." Available from http://www.fas.org/nuke/control/salt1/text/salt1.htm; accessed October 21, 2008; United Nations. 1972. "Treaty Between the United States of America and the Union of Soviet Socialist Republics on the Limitation of Anti-Ballistic Missile Systems."Available from http://www.fas.org/nuke/control/abmt/text/index.html; accessed October 21, 2008 and United Nations. 1979. "Treaty Between The United States of America and the Union of Soviet Socialist Republics on the Limitation Of Strategic Offensive Arms, Together With Agreed Statements and Common Understandings Regarding the Treaty." Available from http://www.fas.org/nuke/control/salt2/text/index.html; accessed October 21, 2008.

9. United Nations. 1987. "Treaty Between the United States of America and the Union of Soviet Socialist Republics on the Elimination of Their Intermediate-Range and Shorter-Range Missiles." Available from http://www.fas.org/nuke/control/inf/text/index.html; accessed October 21, 2008; United Nations. 1991 "Treaty Between the United States of America and the Union of Soviet Socialist Republics on the Reduction and Limitation of Strategic Offensive Arms and Associated Documents." Available from http://www.fas.org/nuke/control/start1/text/index.html; accessed October 21, 2008; and United Nations. 1993. "Treaty Between the United States of America and the Union of Soviet Socialist Republics on the Reduction and Limitation of Strategic Offensive Arms." Available from http://www.fas.org/nuke/control/start2/text/index.html; accessed October 21, 2008.

10. United States Senate, "Moon Treaty," 166.

11. United States House of Representatives. 1979 (September 5 and 6). "International Space Activities, 1979." Hearings Before the Subcommittee on Space Science and Applications, Committee on Science and Technology. 96th Congress, 1st Session. Y4.Sci2:96/50, 101.

12. Ibid., 109–110.

13. Ibid., 110.

14. Ibid., 114.

15. Ibid., 116.

16. Ibid., 101–104.

17. United States Senate, "Moon Treaty," 118.

18. United Nations. 1982. "United Nations Convention on the Law of the Sea." 1833 U.N.T.S. 3. Available from http://www.un.org/Depts/los/convention_agreements/texts/unclos/unclos_e.pdf; accessed November 29, 2004.

19. United States House of Representatives, "International Space Activities," 113.

20. Ibid., 142.

21. United States Senate, "Moon Treaty," 122.

22. United States Senate, "Moon Treaty," 124–130.

23. Rana, Harminderpal Singh. 1994. "The Common Heritage of Mankind and the Final Frontier," *Rutgers Law Review,* 26:225; and Mani, V. S. 1996. "The common heritage of mankind: Implications for the legal status of property rights on the moon and celestial bodies," *Proceedings of the 39th Colloquium on the Law of Outer Space.* American Institute of Aeronautics and Astronautics, 31–37.

24. Ibid., 168.

25. American Bar Association. 1980 (April). "Report to the House of Delegates Section of International Law." Reprinted in United States Senate, "The Moon Treaty," 76–81. Quoted in Finch, Edward R., Jr., and Amanda Lee Moore. "The 1979 Moon Treaty Encourages Space Development." *Proceedings, 23rd Colloquium on the Law of Outer Space.* American Institute of Aeronautics and Astronautics, 1980. 80-SL-06, 13–18.

26. United States Senate, "Moon Treaty," 19–20.

27. Ibid., 143.

28. Ibid., 21.

29. Rosenfield, "A Moon Treaty?"

30. United States House of Representatives, "International Space Activities," 104.

31. United States Senate, "Moon Treaty," 188.

32. Fasan, Ernst. 1980. "Some Legal Problems Regarding the Moon." *Proceedings, 23rd Colloquium on the Law of Outer Space.* American Institute of Aeronautics and Astronautics. 80-SL-05, 9–11.

33. United Nations Committee on the Peaceful Uses of Outer Space. 1972. U.N. Doc. A/AC.105/C.2(XI)WP12.

34. United States Senate, "Moon Treaty," 169.

35. Ibid., 169.

36. Ibid., 170.

37. Ibid.

38. Ibid.

39. Ibid., 123.

40. Ibid., 136.

41. Ibid., 137.

42. United Nations Committee on the Peaceful Uses of Outer Space. 1975. U.N. Doc. A/AC.105/C.2 SR.226–245.

43. United Nations Committee on the Peaceful Uses of Outer Space. 1972b. U.N. Doc. A/AC.105/C.2/SR.188, 32.

44. United Nations Committee on the Peaceful Uses of Outer Space. 1972a. U.N. Doc. A/AC.105/C.2/SR.187, 10–11.

45. Christol, Carl Q. 1980a. "An International Regime, Including Appropriate Procedures, for the Moon: Article 11, Paragraph 5 of the Moon Treaty." *Proceedings, 23rd Colloquium on the Law of Outer Space.* American Institute of Aeronautics and Astronautics. 80-SL-33, 139–148.

46. United States Senate, "Moon Treaty," 24.

47. Ibid., 132.

48. Ibid., 129.

49. Ibid., 66.

50. Stiglitz, Joseph E. 2002 *Globalization and Its Discontents.* New York: W. W. Norton, 11–16.

51. United States Senate, "Moon Treaty," 70.

52. Kennedy, John F. 1961 (January 20). "Inaugural Address." Available from http://www.jfklibrary.org/j012061.htm; accessed January 3, 2006.

53. Danilenko, Gennady M. 1989. "Outer Space and the Multilateral Treaty-Making Process." *Berkeley Technology Law Journal,* 4, 2. Available from http://www.law.berkeley.edu/journals/btlj/articles/vol4/Danilenko/HTML/text.html; accessed March 19, 2005.

54. Dula, Arthur M. 1979. "Free Enterprise and the Proposed Moon Treaty." *Houston Journal of International Law,* 2, 3: 8–9.

55. United States Senate, "Moon Treaty," 24.

56. Haanappel, P. P. C. 1980. "Article XI of the Moon Treaty." *Proceedings, 23rd Colloquium on the Law of Outer Space.* American Institute of Aeronautics and Astronautics, 80-SL-10, 29–33.

57. United States Senate, "Moon Treaty," 120.

58. United Nations Committee on the Peaceful Uses of Outer Space. 1973. U.N. Doc. A/AC.105 PV.123, 6; United Nations Committee on the Peaceful Uses of Outer Space. 1974. U.N. Doc. A/AC.105 C.2 SR.211, 26; and United Nations Committee on the Peaceful Uses of Outer Space. 1976. U.N. Doc. A/AC.105 C.2 SR.249, 8.

59. United Nations Committee on the Peaceful Uses of Outer Space. 1973b. A/AC.105/C.2/SR.205, 115.

60. Finch, Edward R., Jr., and Amanda Lee Moore. 1980. "The 1979 Moon Treaty Encourages Space Development." *Proceedings, 23rd Colloquium on the Law of Outer Space.* American Institute of Aeronautics and Astronautics, 80-SL-06, 13–18.

61. Rosenfield, "A Moon Treaty?"

62. United Nations Committee on the Peaceful Uses of Outer Space. 1979. U.N. Doc. A/AC.105/PV.203, 22; and United States House of Representatives, "International Space Activities," 86.

63. United States House of Representatives, "International Space Activities," 86.

64. Ibid., 115.

65. Dula, "Free Enterprise," 18–19.

66. Galloway, Eilene. 1980. "Issues in Implementing the Agreement Governing the Activities of States on the Moon and Other Celestial Bodies." *Proceedings, 23rd Colloquium on the Law of Outer Space.* American Institute of Aeronautics and Astronautics, 80-SL-08, 19–24.

67. United States Senate, "Moon Treaty," 22.

68. United Nations Committee on the Peaceful Uses of Outer Space. 1979. U.N. Doc. A/34/20; and United States House of Representatives, "International Space Activities," 86.

69. Dula, "Free Enterprise," 10.

70. Ibid., 8.

71. Greenspan, Alan. 1996 (December 5). "The Challenge of Central Banking in a Democratic Society." Remarks at the Annual Dinner and Francis Boyer Lecture of The American Enterprise Institute for Public Policy Research, Washington, D.C. Available from http://www.federalreserve.gov/boardDocs/speeches/1996/19961205.htm; accessed January 31, 2006.

72. United Nations Committee on the Peaceful Uses of Outer Space. 1973. U.N. Doc. A/AC.105 PV.123, 6.

73. Dula, "Free Enterprise," 12.

74. van Traa-Engelman, H. L. 1980. "The Moon Treaty: Legal Consequences and Practical Aspects." *Proceedings, 23rd Colloquium on the Law of Outer Space.* American Institute of Aeronautics and Astronautics, 80-SL-53.73–77.

75. Haanappel, "Article XI."

76. United Nations Committee on the Peaceful Uses of Outer Space. 1972a. U.N. Doc. A/AC.105/C.2/SR.187, 10–11.

77. Dekanozov, R. V. 1980. "Juridicial Nature and Status of the Resources of the Moon and Other Celestial Bodies." *Proceedings, 23rd Colloquium on the Law of Outer Space.* American Institute of Aeronautics and Astronautics. 80-SL-03, 5–8.

78. Rosenfield, "A Moon Treaty?"

79. Ibid.

80. Menter, "Commercial Space Activities."

81. United Nations Committee on the Peaceful Uses of Outer Space. 1979. U.N. Doc. A/34/20; and United States House of Representatives, "International Space Activities," 86.

82. Okolie, Charles Chukwuma. 1980. "Legal Interpretation of the 1979 United Nations Treaty Concerning the Activities of Sovereign States of the Moon and Other Celestial Bodies Within the Meaning of the Concept of the Common Heritage of Mankind." *Proceedings, 23rd Colloquium on the Law of Outer Space.* American Institute of Aeronautics and Astronautics, 80-SL-58, 61–67.

83. United States Senate, "Moon Treaty," 132.

84. Ibid.

85. Ibid., 50.

86. Ibid., 42–43.

87. Wadegaonkar, Damodar. 1984, *Orbit of Outer Space Law*. London: Stevens and Sons, 37.

88. Ishinomori, Shotaro. 1988. *Japan, Inc.* Berkeley: University of California Press.

89. Dula, "Free Enterprise," 20.

90. Finch and Moore, "The 1979 Moon Treaty," 13–18.

91. Haanappel, "Article XI."

92. Galloway, "Issues in Implementing the Agreement."

93. Menter, "Commercial Space Activities."

94. Heinlein, Robert A. 1966. *The Moon Is a Harsh Mistress*. New York: G. P. Putnam.

95. Center for International Environmental Law. 2004. "Groups Defend California's Right to Protect Public Health: Canadian Corporation's NAFTA Suit Threatens State Sovereignty." Available from http://www.ciel.org/Tae/Methanex_30Mar04.html; accessed December 30, 2005.

96. International Criminal Tribunal for the Former Yugoslavia. 2002. "The Trial of Slobodan Milosevic." Available from http://www.un.org/icty/milosevic/; accessed December 28, 2005.

97. Dula, "Free Enterprise," 14.

98. United States Senate, "Moon Treaty," 52.

99. Dula, "Free Enterprixe," 14.

6

Return to the Moon: The Moon Agreement Reconsidered

AN AUTHORITATIVE INTERPRETATION

The final arguments in favor of the Moon Agreement rest on the question of which of the several interpretations of the document are most valid. As Articles 31 and 32 of the Law of Treaties requires, if a clear interpretation is not found in the agreement itself, there is recourse "to supplementary means of interpretation, including the preparatory work of the treaty and the circumstances of its conclusion." No opponent of the Moon Agreement has used "the preparatory work" in their arguments, whereas many proponents have. This alone strongly suggests that the proponents are on much better legal ground. However, what we have not yet examined are "the circumstances of its conclusion."

We turn again to Robert B. Owen, State Department legal adviser, and the answers he gave to the questions the Senate Space Subcommittee sent him after his testimony:

Question 6: Before reaching consensus on the agreement, the Committee on the Peaceful Uses of Outer Space made a decision to clarify certain parts of the agreement by stating certain understandings in its report. There are three understandings to be found in paragraphs 62, 63, and 65 of the Committee's report. It is noted that the resolution adopted by the General Assembly endorsing the treaty called attention to these three paragraphs. What is the legal status of these three understandings?

Answer: Under the customary international law of treaty interpretation, as codified in Articles 31 and 32 of the Vienna Convention on the Law of Treaties, the understandings set out in paragraphs 62, 63, and 65 of the 1979 Report of the Committee on the Peaceful Uses of Outer Space constitute an authoritative interpretation of the questions they address.[1]

Indeed, one finds that General Assembly Resolution 34/68, dated December 5, 1979 (see Appendix 5), which endorsed the Moon Agreement, includes the clause:

> *Having considered* the relevant part of the report of the Committee on the Peaceful Uses of Outer Space, in particular paragraphs 62, 63 and 65.[2]

Nearly all discussion of the Moon Agreement overlooks the understandings in the 1979 COPUOS report, yet these paragraphs are essentially part and parcel of the agreement, as much as if they appeared in the body of the agreement itself (see Appendix 6). They are the committee's official, unanimous interpretation of key provisions of the treaty that they themselves drafted. There simply is no higher source of interpretation, since the source of the interpretation of the agreement is the source of the agreement itself. This is especially true in the absence of any understandings to the agreement promulgated by the United States. In fact, the COPUOS understandings probably fall under the provisions of the Vienna Convention on the Law of Treaties, Article 31, paragraph 3(a), that "[t]here shall be taken into account, together with the context" of the treaty:

> Any subsequent agreement between the parties regarding the interpretation of the treaty or the application of its provisions.[3]

Under this theory, the COPUOS understandings carry greater weight than documents that comprise the "preparatory work of the treaty" to be considered under Article 32. Eilene Galloway's account of how the COPUOS understandings came into existence supports this theory:

> An unusual procedure was adopted by the Committee on the Peaceful Uses of Outer Space at [the end of the negotiation]. Certain of the draft treaty articles required interpretation and instead of amending the text, Committee understandings were set forth in the Committee's report, and these understandings were officially included by the General Assembly by reference when adopting the treaty text. The reason for this type of action was that consensus could be achieved on a treaty test prepared by Austria and already studied by the delegates, but if it were subjected to many possible changes emerging consensus would be imperiled.[4]

Thus, the understandings were a more convenient mechanism for completing the committee's work on the Moon Agreement than inserting the language in the body of the agreement, but the understandings are de facto an annex to the agreement. Any reading of the Moon Agreement that doesn't consider the COPUOS understandings is as erroneous as a reading of the Constitution of the United States that doesn't consider the

Bill of Rights, or all of its amendments. General Assembly Resolution 34/68 captures the agreement and the COPUOS understandings as a package. Although the understandings do not lay to rest all of the controversies, they deliver the final word on three of them. Paragraph 65 addresses the alleged mining moratorium:

> Following a suggestion for further clarification of article VII, the committee agreed that article VII is not intended to result in prohibiting the exploitation of natural resources which may be found on celestial bodies other than the Earth but, rather, that such exploitation will be carried out in such manner as to minimize any disruption or adverse effects to the existing balance of the environment.[5]

This slays the biggest bogeyman in the nightmares of the Moon Agreement opponents. The tragedy is that the mechanism that was expedient for finalizing the Moon Agreement has caused its authoritative interpretation to be overlooked.

REACHING FOR UNDERSTANDING

The positivist school of international law considers norms to be defined by states; the natural law school seeks to derive norms from basic metaphysical principles. Strong states favor positivism: they do only what they want to do and are bound only by that which they choose. Unsurprisingly, the United States is highly positivist in its approach to international law. There are two mechanisms by which states ensure that the treaties to which they become party apply to them only in ways they choose: reservations and understandings.

A reservation is a declaration that a state makes to exclude or alter the legal effect of certain provisions of the treaty in their application to that state. A reservation enables a state to accept a multilateral treaty as a whole by allowing it to avoid compliance with certain provisions. Reservations can be made when the treaty is signed, ratified, accepted, approved, or acceded to. Reservations must not be incompatible with the object and the purpose of the treaty. Furthermore, a treaty might prohibit or limit reservations.*

Sometimes states make declarations as to their understanding of some matter or as to the interpretation of a particular provision. Unlike reservations, declarations of understanding merely clarify the state's position

---

* A state can make a reservation, but another state can object and even refuse to regard the reserving state as party to the treaty.

and do not purport to exclude or modify the legal effect of a treaty. Usually, such declarations are made at the time of the deposit of the corresponding instrument or at the time of signature.[6]

Thus it would be perfectly legitimate for the United States to become party to the Moon Agreement and declare reservations or understandings as the government sees fit to protect American interests, pursuant to Articles 19 through 23 of the Vienna Convention on the Law of Treaties. Ronald F. Stowe, chairman of the Aerospace Law Committee, Section of International Law, American Bar Association, opines:

> I do not believe that any of these general purposes for the international regime are the least bit, by themselves, offensive to U.S. interests, either now or in the future. I suggest rather that a hard-nosed, skeptical, self-interested analysis leads to the conclusion that ratification of the treaty subject to the inclusion in our instrument of ratification of certain specific national interpretations of key issues would put the United States in a far stronger position to protect out interests, both legally and politically, than would refusal to ratify.
>
> The inclusion of interpretations in our instrument of ratification would legally define the scope of the obligations which we have assumed, and no other country would have a defensible claim that we are bound by any other interpretation.
>
> It should come as no surprise that countries with a present or anticipated ability to engage in lunar resource exploitation will seek maximum flexibility to do so. That is what we are doing. That is exactly what we should do.
>
> I find a curious irony in the recommendation by some that the best way to protect our national interest is not only to refuse to play the game, but also to refuse to try to shape the rules.
>
> Why refuse to shape the rules? Because we anticipate the majority will not agree with us? That recommendation appears to presume and project a national impotence, lack of will, absence of strength of our convictions and inability or unwillingness to protect ourselves.[7]

In 1980, the report of the Section of International Law of the American Bar Association included four sample understandings with regard to the Moon Agreement (see Appendix 7):

> [N]o provision in this Agreement constrains the existing right of governmental or authorized non-governmental entities to explore and use the resources of the moon or other celestial body, including the right to develop and exploit these resources for commercial or other purposes. . . . nothing in this Agreement in any way diminishes or alters the right of the United States to determine how it shares the benefits derived from exploitation by or under the authority of the United States of natural resources of the moon or other celestial bodies.

Natural resources extracted, removed or actually utilized by or under the authority of a State Party to this Agreement are subject to the exclusive control of, and may be considered as the property of, the State Party or other entity responsible for their extraction, removal or utilization.

The meaning of the term "common heritage of mankind" is to be based on the provisions of this Agreement, and not on the use or interpretation of that term in any other context. . . . [Acceptance] constitutes recognition that all States have equal rights to explore and use the moon and its natural resources. . . . States Parties retain exclusive jurisdiction and control over their facilities, stations and installations on the moon, and that other States Parties are obligated to avoid interference with normal operations on such facilities.

Acceptance . . . in no way prejudices the existing right of the United States to exploit or authorize the exploitation of those natural resources. No moratorium on such exploitation is intended or required by this Agreement. . . . the United States reserves to itself the right and authority to determine the standards for [acting in a manner compatible with the provisions of the agreement] unless and until the United States becomes a party to a future resources exploitation regime. In addition, acceptance of the obligation to join in good faith negotiation of such a regime in no way constitutes acceptance of any particular provisions which may be included in such a regime; nor does it constitute an obligation to become a Party to such a regime regardless of its contents.[8]

What more guarantee and wiggle room could anyone in favor of the commercial development of outer space possibly want? Carl Q. Christol, international law and political science professor, welcomes these understandings for providing political clarity. He also considers them unnecessary for legal clarity to the Moon Agreement:

These, in my view can be justified on the ground both as an aid to clarity and as excessive caution. . . . I have concluded that the proposed "Understandings" in no way modify the terms of the treaty. They are welcome. But they are not necessary since what they contain can be found within the treaty upon a close and careful reading of it taking into account, as I have, the negotiating history of the Agreement.[9]

Another option for making the agreement fit U.S. needs is to revise the agreement before the United States becomes a party to it. Article 18 provides:

Ten years after the entry into force of this Agreement, the question of the review of the Agreement shall be included in the provisional agenda of the General Assembly of the United Nations in order to consider, in the light of past application of the Agreement, whether it requires revision.

The Moon Agreement completed its tenth year in force in 1994, thus the door is open to revise the agreement.

AWAITING THE NEW MOON

Twenty years ago, international law expert Gennady M. Danilenko observed:

> Many states failed to ratify the Moon Treaty because they felt that it was premature. Indeed, one can hardly claim that there is pressing need to adopt legal rules at this stage which purport to govern mini-activities on the moon and other celestial bodies. Such activities will take place only in the very distant future.[10]

In the first decade of the twenty-first century, as the United States gears up to return to the Moon, as Russia considers plans to establish a helium-3 mining facility on the Moon, and both China and India consider their own manned lunar exploration programs, is Danilenko's statement still true? Is it possible that the Moon Agreement is at last becoming necessary, or is the less specific language of the Outer Space Treaty sufficient? The legality of resource appropriation (as opposed to real property rights) can be—and has been—derived indirectly from the Outer Space Treaty; however, as discussed in chapter five, "Resource Property Rights," the Moon Agreement specifies the right to acquire ownership of extracted natural resources under the provisions of Article 11, paragraph 3. Other specific issues can be addressed in the context of the Moon Regime. For example, an issue raised in author Wayne N. White's concept of functional property rights, namely, how large a zone of exclusive use should an entity be entitled to in order to ensure freedom from interference by another entity's operation, is one of the standards that could be established by the Moon Regime.[11] As another example, Edward R. Finch, Jr. and Amanda Lee Moore, along with Danilenko, point out that the environmental protections in the Moon Agreement are stronger and more specific than those on the Outer Space Treaty:

> [T]he potential benefits to the U.S. of ratification of the Moon Treaty significantly outweighs any potential disadvantages. This is particularly true in light of the fact that such potential disadvantages would arise not from the Moon Treaty, but rather from a subsequent accord which will require separate approval of disapproval. Refusal to ratify would simply and devastatingly exclude the U.S. from legal entitlement to the benefits and protections

included in the Moon Treaty, and could seriously complicate the prospects of gaining concessions from other States in the future regime negotiations. In itself, this would only give rise to investor insecurity and probably make it more difficult for the U.S. to shape the results of those future negotiations.[12]

A comparative analysis of Article IX of the Outer Space Treaty and of Article 7 of the Moon Treaty indicates that the content of the obligations relating to the protection of the earth and space environments imposed by these treaties is different. Regarding outer space, Article IX of the Outer Space Treaty expressly limits the relevant environmental obligations to activities relating to the "study" and "exploration" of outer space. Other types of space activities, including such environmentally significant activities as the exploitation of the resources of outer space, do not seem to fall within the purview of Article IX. With respect to the earth, Article IX requires only the avoidance of environmental hazards relating to the possible introduction of extraterrestrial matter. It does not contain a general environmental obligation applicable to all space activities. In contrast, the environmental protection rules of the Moon Treaty cover all possible kinds of adverse effects on the moon's environment, as well as the earth's, which may result from activities associated with the exploration and use of the moon and other celestial bodies.[13]

During hearings on the Moon Agreement in the U.S. Senate, Harrison Schmitt (R-NM) asked:

> How can we properly negotiate and then examine and ratify a treaty of this kind when we have no space policy with respect to exploitation of either inner space or outer space?
>
> We have no policy. There was no statement we will exploit weightlessness and vacuum for commercial or noncommercial purposes. No policy statement that we will move forward, in some kind of realistic way, the further exploration of the Moon and of other planets and potential exploitation of the resources of the same. We have no such policy. If we do, I haven't heard about it. Our policy is that we don't do that, in fact. It has been a Presidential statement.
>
> We are in a technological position to proceed. Nobody knowledgeable about this area doesn't believe that. Just because you can't do it tomorrow doesn't mean you are not in a technological position to proceed.[14]

Since Senator Schmitt made that statement in 1980, the United States has been "in a technological position to proceed" to return to the Moon and to establish a permanent presence there. What it lacked during the past quarter century is the will to proceed. However, perhaps the statement of President George W. Bush on January 14, 2004 signaled a change:

> Our third goal is to return to the moon by 2020, as the launching point for missions beyond. Beginning no later than 2008, we will send a series of

That was a lie

robotic missions to the lunar surface to research and prepare for future human exploration. Using the Crew Exploration Vehicle, we will undertake extended human missions to the moon as early as 2015, with the goal of living and working there for increasingly extended periods. . . .

Returning to the moon is an important step for our space program. Establishing an extended human presence on the moon could vastly reduce the costs of further space exploration, making possible ever more ambitious missions. Lifting heavy spacecraft and fuel out of the Earth's gravity is expensive. Spacecraft assembled and provisioned on the moon could escape its far lower gravity using far less energy, and thus, far less cost. Also, the moon is home to abundant resources. Its soil contains raw materials that might be harvested and processed into rocket fuel or breathable air. We can use our time on the moon to develop and test new approaches and technologies and systems that will allow us to function in other, more challenging environments. The moon is a logical step toward further progress and achievement.

With the experience and knowledge gained on the moon, we will then be ready to take the next steps of space exploration: human missions to Mars and to worlds beyond.[15]

The President's Commission on Implementation of United States Space Exploration Policy, chaired by former Secretary of the Air Force Edward C. Aldridge, states:

Over time, missions to the Moon, Mars, and beyond will test various methods for finding commercial value in space, including use of in situ or space resources. Collecting and transmitting energy to re-power satellites, mining mineral resources, conducting new materials research, or low gravity manufacturing: the public advanced these and many more ideas to us.[16]

On December 22, 2005, the Senate passed the FY 2006 NASA Authorization Bill, which included full funding for the "Vision for Space Exploration." For FY 2009, Congress authorized several billion dollars more for NASA than the Bush administration requested.[17] Despite the global financial crisis that erupted during the transition to the Obama administration, President Barack Obama has committed himself to keeping NASA's Constellation program on track to return astronauts to the Moon by 2020.[18] So far, so good. Indeed, it appears that the space policy of the United States is that it shall proceed to the Moon and to other celestial bodies, and that this effort will include commercial exploitation of their natural resources. Thus many of the concerns that Schmitt initially raised about the timing of consideration of the Moon Agreement have been removed.

Meanwhile, other launching states have their eyes on the Moon. Luan Enjie, vice-minister of China's Commission of Science, Technology and

Industry for National Defense and director of the China National Aerospace Administration (CNAA), outlines China's lunar plans in an interview published by *People's Daily*. These plans include developing the Moon's helium-3 as an energy resource:

> The Moon has become the focal point wherein future aerospace powers contend for strategic resources. The Moon contains various special resources for humanity to develop and use. [Helium-3] is a clean, efficient, safe and cheap new-type nuclear fusion fuel for mankind's future long-term use, and it will help change the energy structure of human society.[19]

Nikolai Sevastyanov, former cosmonaut and current director of RKK Energiya, has stated that Russia is planning to mine helium-3 on the Moon by 2020 at a permanent base supported by a heavy-cargo transportation system:

> We are planning to build a permanent base on the Moon by 2015 and by 2020 we can begin the industrial-scale delivery . . . of the rare isotope helium-3.[20]

Schmitt, who now pilots the research firm Interlune-Intermars Initiative, Inc., has said:

> There's enough in the Mare Tranquillitatis alone to last for several hundred years.[21]

The isotope is rare on Earth, but has been deposited on the lunar surface in quantity over billions of years by the solar wind, the high-speed stream of charged particles flowing from the sun. Fusion research has been ongoing since the mid-twentieth century, first as a weapon of mass destruction, followed by interest in its possibilities as a means of power generation. The initial reaction of interest was the fusion of hydrogen-2 (deuterium) and hydrogen-3 (tritium) to form helium-4 and expel a high-speed neutron:

$$^2_1D + {^3_1}T \rightarrow {^4_2}He \ (3.5 \text{ MeV}) + n^0 \ (14.1 \text{ MeV})$$

The excess neutron constitutes a source of intense particle radiation. This is not particularly a concern in a thermonuclear weapon, where the intent is to inflict damage, and is even desirable in applications where the preferred kill mechanism is enhanced radiation versus heat and blast damage (the so-called neutron bomb). However, if the D-T reaction were to be sustained and controlled as a power source, it would necessitate heavy shielding; the excess neutron, having no electric charge, is not subject to being controlled by a magnetic field. Over time, as the shielding material absorbs neutrons, its nonradioactive nuclides are transmuted into radioactive ones, creating a waste problem. Some researchers have settled on helium-3 as a more practical fusion reaction for power generation:

$$^2_1D + {}^3_2He \rightarrow {}^4_2He\ (3.6\ \text{MeV}) + p^+\ (14.7\ \text{MeV})$$

The reaction produces a helium-4 nucleus, which is nonradioactive, and a proton, which is positively charged and easily channeled by a magnetic field into useful energy.[22]

The problem is that only a few hundred kilograms of helium-3 are known to exist on Earth. Early indications are that the isotope is more plentiful on the lunar surface, although some skeptics question whether it exists in concentrations that would be commercially feasible to extract. Another barrier to commercial feasibility is getting more energy out of the reaction than the energy necessary to create and control it. Fusion only occurs at temperatures of several million degrees Celsius, an environment in which the electrostatic repulsion of nuclei is overcome and collisions can occur (claims for cold fusion reactions have not been reproduced). Thirty years ago, it was said that commercial fusion power generation was twenty to fifty years in the future; some say this is still a good estimate. So, a new race for the Moon is being touted in the press, although it remains to be seen whether any vigorous Moon effort, either born of competition or of cooperation, is on the horizon.

In considering the entirety of the Moon Agreement, its negotiation history, and particularly the understandings in the 1979 COPUOS report, the conclusion is inescapable that there was never any legal justification for rejecting the agreement. Few of the opponents of the agreement refer to its negotiation history, and those who do deceptively characterize the uncontradicted statements of the United States in COPUOS as "unilateral interpretations," whereas they constitute the highest source of interpretation outside of the text of the agreement itself. Thus in all cases the opponents to the agreement offer only flawed, incomplete, and biased analyses of the agreement.

In considering the testimony offered against the treaty, the conclusion is inescapable that the agreement was defeated not on its own merits, but as a prelude to defeating the UNCLOS III as part of a larger political-economic conflict between the United States and the Third World at a time when, in the aftermath of the Vietnam disaster as well as ongoing hostage crises in Teheran and Beirut, growing Soviet influence in the Horn of Africa and Angola, and the Soviet invasion of Afghanistan, the American psyche was severely traumatized vis-à-vis its relations with the rest of the world. It was a United States that had lost its New Frontier optimism; it was a United States that trembled to say "yes." The sentiments expressed by George S. Robinson, legal counsel at the Smithsonian Institution, and Harold M. White Jr., executive vice president at the University of West Florida, are from a bygone era of Cold War paranoia:

> The Soviets, after long opposing the insistence, primarily of the United States, on the principle of *res communis* in outer space, did a little more research into the breadth of Western legal thinking and into the value of third world voting power for the eventual establishment of an international law of peaceful coexistence. Their tactics in the Moon Treaty negotiations, both during and after completion, were to convince the small nations of the benefit to be derived from redefining the traditional meaning of *res communis*. . . .
>
> It should not be suggested, however, that the Soviets have craftily manipulated the third world. . . . The Soviets have rather influenced and benefited from underlying trends that already existed. The change in Soviet thinking on *res communis* can really be traced to opportunities presented by Western objectives in the law of the sea negotiations, by third world objectives in negotiations for new information and economic orders, and in phrases contained in the Outer Space Treaty concerning space as the "province of mankind" and "for the benefit of mankind."[23]

Here Robinson and White also refer to anonymous remarks "by an experienced Washington, D.C., attorney involved in helping identify the interests of the private sector in the resources of the ocean and outer space." With some degree of confidence we can identify the attorney as Leigh Ratiner, but we may wonder why he chose to remain anonymous.

In any case, Robinson and White warn of the Soviets' ability to use to their own advantage the high ideals expressed in U.S. foreign policy. In a similar vein, author John A. Stormer suggested a generation earlier not that U.S. interests were being betrayed by a deliberate, fifth-column, Communist conspiracy, but by a "conspiracy of shared values" among U.S. elites:

> Is there a conspiratorial plan to destroy the United States into which foreign aid, planned inflation, distortion of treaty-making powers and disarmament all fit? . . . Could some of those who make the tragic decisions and implement the wrong programs consistently be Communists? . . . We don't know and we can't know.[24]

Stowe, speaking before the Senate Space Subcommittee in 1980:

> We have all heard the current rash of horror stories about the consequences of ratification. For example, we have heard that under this treaty the United States will have to give away its advanced technology to all comers. We have heard the international community can seize the benefits of U.S. efforts in lunar research exploitation. Even though negotiators rejected it and the

treaty does not mention it, we have heard that there is somehow created a moratorium on such exploitation.

We have heard that foreign countries could, at their whim, send storm troopers marching into our space facilities; that encroachment on space operation is permitted, if not encouraged, that the treaty would bind and constrain the United States but not other states parties and a variety of other such improbables.

I believe these stories are not based on what the treaty says, but are principally the products of excited imaginations combined with generalized frustration with the current ineffectiveness of the United States in foreign affairs. This combination appears to lead to a blanket assumption that we have been had, regardless of what the treaty actually says. That frustration may be understandable, but it is hardly a promising or adequate foundation on which to build our inescapable future role in the international community.[25]

That was over a quarter century ago. With the exceptions of Cuba and North Korea, the former Communist Bloc has transitioned to capitalism, and the free-trade paradigm of globalization has swept the Third World. The alleged socialist intent of the "common heritage" concept has been deprived of its theoretical underpinnings. Since the era of the New International Economic Order (NIEO), the debate over the structure of the global economy has shifted from one about central planning versus markets to one about what type of market economy, for example, moderate government intervention along Keynesian principles, the "developmental state" strategies of the Asian "tiger" economies, and laissez-faire neoliberalism. If indeed "the march of freedom and democracy" has left "Marxism-Leninism on the ash heap of history," as Ronald Reagan predicted it someday would, what does the United States, as the sole superpower in a capitalist world, have to fear from a clearly erroneous socialist interpretation of the Moon Agreement?[26] If the United States is now serious about returning to the Moon, and this time to stay, which necessarily means exploiting its natural resources to support permanent lunar bases, then the time is coming soon when debate on the Moon Agreement should be reopened in the State Department and ultimately in the Congress. A spirited public debate on the merits of the agreement might contribute to heightened public interest in the nation's effort to return to the Moon to stay, and to go to Mars and beyond.

In 1980, Ratiner listed the conditions under which the L-5 Society would have supported the Moon Agreement:

Mr. Chairman, the L-5 Society believes that United States interests would benefit from a treaty on the resources of the moon and other celestial bodies which is designed to promote exploration and exploitation and at the same time protect the safety of human life and the environment once industrial activity occurs in outer space; which ensures that countries respect their reciprocal rights to extract resources; and which guarantees that countries do not claim resource deposits as a matter of hegemony but rather restrict their claims in direct relation to their capabilities to extract the resources and use them. Indeed, some have argued that the United States interpretation of the common heritage concept embodies these same principles. If these concepts, together with the principle of free access, were to represent the agreed guidance provided in the Moon Treaty for the subsequent regime, the L-5 Society would be supporting United States ratification.[27]

This chapter has shown that the Moon Agreement indeed is and does all of these things, when considered in the light of the authoritative interpretation, as expressed by the very committee that crafted the document. The only interpretation of the "common heritage" concept that can have any legal meaning is that it "finds its expression in the provisions of this Agreement," is thus merely hortatory, and has no connection with whatever meaning the term might have in the Law of the Sea. Thus the L-5 Society should have supported United States ratification of the Moon Agreement in 1979, and there is no reason why its successor organization, the National Space Society, should not support ratification now.

As expressed in 1980 in a letter to the Senate Space Subcommittee, the Aerospace Industries Association's (AIA) opposition was conditional:

We urge that prior to signature and ratification by the United States, the following set of principles be clearly embodied in the Treaty:

(a) The unilateral right of exploitation prior to the establishment of an international regime;
(b) The protection of the participants' intellectual property on a commercially acceptable basis;
(c) The reservation to the participant of a commercially acceptable portion of the benefit of its exploitation;
(d) The immediate undertaking to establish the international regime promptly;
(e) Clear limitations on the authority of the regime consistent with the foregoing and in accordance with principles which will protect the commercial investment and participation of private enterprise in the exploitation of the moon.[28]

This chapter has shown that the Moon Agreement in its present form satisfies points (a) and (b), and points (c) and (e) would unquestionably be the basis on which the United States would negotiate the establishment of the Moon Regime. Additionally, provisions of the Moon Agreement must now be considered in the light of international law that has developed since it was opened for signature in 1979. As examples, the 1996 Declaration on International Cooperation in the Exploration and Use of Outer Space for the Benefit and in the Interest of All States, Taking into Particular Account the Needs of Developing Countries (Declaration on International Cooperation) reflects more relaxed principles regarding the sharing of benefits than appeared in earlier documents, and international intellectual property law developed in the Uruguay Round of the General Agreement on Tariffs and Trade presumably would apply in outer space as well as on Earth.[29] As for point (d), although opinions vary as to whether a state must be party to the Moon Agreement in order to be eligible to participate in the negotiation of the Moon Regime, there certainly would be no better way to begin this process promptly and to ensure a seat at the table than to sign and ratify the Moon Agreement. Under these conditions, the AIA should now support ratification. The American Institute of Aeronautics and Astronautics, which took no stand in 1980, should take a positive stand now.

In 1999, Declan J. O'Donnell, President of United Societies in Space, Inc., as well as then-president of the World Space Bar Association, writes:

[A] tension exists between industry plans for space development and national standards for treaty compliance. Space resource utilization is at the center of that problem.

A possible solution may be found in the Moon Treaty. It can be read to approve space resource utilization for non-commercial purposes, i.e. construction of habitat on the Moon. It appears to approve commercial uses, also, subject to the treaty burden of equitable benefit sharing. The "equitable" standard is one that court systems worldwide apply on a day to day basis without much legal difficulty. The lingering problem with this treaty is that a new concept was also mentioned, the common heritage of mankind. Thus, the benefit of this treaty could be canceled by the burden of this new concept.[30]

Where does O'Donnell claim to find the definition of "the common heritage of mankind is defined"? In the Law of the Sea Convention. However, he does not cite article and paragraph, nor can he, it would seem, since Article 136 of the convention, "Common Heritage of Mankind," states *in toto*, "The Area and its resources are the common heritage of mankind." This is a declarative statement, not a definition, but if the convention contained a definition, one would certainly expect to find it in Article 136. In any case, as has been demonstrated, it is simply erroneous to apply the UNCLOS III to the Moon Agreement. Disabused of this misconception, and noting that the 1996 Declaration on International Cooperation takes the starch out the "benefit sharing" concept, one can surmise that O'Donnell would be compelled to support the Moon Agreement wholeheartedly:*

> Another avenue of solution for those who care to develop the Moon, Mars, and all orbits in space may be found in another provision in the Moon Treaty. This is the provision that calls for adoption of another legal regime when space development appears to be feasible. Obviously, this opportunity should be seized by industry leaders to propose a legal regime and a space governance paradigm that encourages development.[31]

O'Donnell notes that in the implementing legislation in Congress, the United States required that it be the managing partner of the Sea Bed Authority:

> Consent of the other nations was forthcoming so that treaty was adopted and the seabed is being mined under congressional implementing legislation by an authority.[32]

It took a quarter of a century, but in the end the United States achieved a satisfactory outcome for its interests in the Law of the Sea Convention. Certainly, with regard to the Moon Agreement, this same perseverance could achieve a similarly happy ending.

Although she erroneously blames the paralysis in space development on the "common heritage of mankind" principle, and gives a shockingly superficial treatment of the Outer Space Treaty and the Moon Agreement (which she assumes to be the root of the problem), lawyer Lynn M. Fountain nevertheless manages to find a solution:

> A free-market approach bolstered by the legal certainty inherent in a system that provides defined property rights would do much to energize the stalled development of the space industry. Involvement of private companies can provide the focus, money, and research necessary for successful growth. But such growth must take place under an international regulatory regime. If the space powers each create and pursue their own legal systems for the commercialization of outer space, the result will be chaotic and prone to international conflict. In addition, involvement of the developing states would be minimal, if present at all. Such a structure would fail to provide the predictable and stable environment necessary for private industry's involvement, and would fail to win international approval and acceptance.

---

* The declaration contains only vague blandishments regarding "the needs of developing countries" (Paragraph 1). However, Paragraph 2 leaves states "free to determine all aspects of their participation in international cooperation in the exploration and use of outer space on an equitable and mutually acceptable basis." Furthermore, there is a strong statement regarding intellectual property rights.

The regulatory model should incorporate the underlying principles of the Outer Space Treaty, while providing a stable basis for private industry's investment in outer space. . . .

The chief purpose of this international regulatory agency would be to create an equitable, efficient, and stable legal environment for the commercial development of outer space. The agency should promote cooperation and opportunity for all interested states, as well as for private industry. The specific purposes of this agency would be: to promote productivity, investment, and the development of expertise; to protect property rights and profits; to promote international cooperation; to develop a set of affirmative duties; and to provide a forum for dispute resolution.[33]

This solution can also be found in what Fountain assumes is the problem: the 1979 Moon Agreement. There is no reason, on Earth or anywhere else in the solar system, why the regime envisioned in the agreement cannot implement these provisions. Fountain has sound ideas regarding the shape that the regulatory regime should take, and there are other good ideas out there. Interested parties should be debating these ideas and working toward consensus. However, it is pusillanimous to shrink from principles because the devil might be in the details. It is time that the United States ratified the Moon Agreement, so that the international community can follow suit and in due course get on with negotiating Moon Agreement II, which will define and establish the regulatory regime.

NOTES

1. United States Senate. 1980. "The Moon Treaty." Hearings Before the Subcommittee on Science, Technology, and Space of the Committee on Commerce, Science, and Transportation. 96th Congress, 2nd Session, 22.

2. United Nations. 1979. "Resolution Adopted by the General Assembly 34/68: Agreement Governing the Activities of States on the Moon and Other Celestial Bodies." Available from http://www.oosa.unvienna.org/SpaceLaw/gares/html/gares_34_0068.html; accessed January 5, 2006.

3. United Nations. 1969. "Vienna Convention on the Law of Treaties." 1155 U.N.T.S. 331. Available from http://www.amanjordan.org/english/un&re/un2.htm; accessed July 1, 2005.

4. United States Senate, "Moon Treaty," 175.

5. United Nations Committee on the Peaceful Uses of Outer Space. 1979. U.N. Doc. A/34/20.

6. United Nations. 2005. "Oceans and Law of the Sea: Chronological Lists of Ratifications of, Accessions and Successions to the Convention and the Related

Agreements as at 20 September 2005." Available from http://www.un.org/Depts/los/reference_files/chronological_lists_of_ratifications.htm#Agreement%20relating%20to%20the%20implementation%20of%20Part%20XI%20of%20the%20Convention; accessed December 24, 2005.

7. United States Senate, "Moon Treaty," 68–69.

8. American Bar Association. 1980 (April). "Report to the House of Delegates Section of International Law." Reprinted in United States Senate, "Moon Treaty," 76–81. Quoted in Finch, Edward R., Jr., and Amanda Lee Moore. "The 1979 Moon Treaty Encourages Space Development." *Proceedings, 23rd Colloquium on the Law of Outer Space.* American Institute of Aeronautics and Astronautics, 1980. 80-SL-06, 13–18.

9. United States Senate, "Moon Treaty," 185.

10. Danilenko, Gennady M. 1989. "Outer Space and the Multilateral Treaty-Making Process." *Berkeley Technology Law Journal*, 4, 2. Available from http://www.law.berkeley.edu/journals/btlj/articles/vol4/Danilenko/HTML/text.html; accessed March 19, 2005.

11. White, Wayne N. 1998. "Real Property Rights in Outer Space." *Proceedings, 40th Colloquium on the Law of Outer Space.* American Institute of Aeronautics and Astronautics, 370. Available from http://www.spacefuture.com/archive/real_property_rights_in_outer_space.shtml; accessed March 19, 2005.

12. Finch, Edward R., Jr., and Amanda Lee Moore. 1980. "The 1979 Moon Treaty Encourages Space Development." *Proceedings, 23rd Colloquium on the Law of Outer Space.* American Institute of Aeronautics and Astronautics, 80-SL-06, 13–18.

13. Danilenko, "Outer Space."

14. United States Senate, "Moon Treaty," 32–33.

15. Bush, George W. 2004. "A New Vision for the Space Exploration Program." Available from http://www.whitehouse.gov/news/releases/2004/01/20040114-3.html; accessed January 6, 2006.

16. Aldridge, Edward C., Carlton S. Fiorina, Michael P. Jackson, Laurie A. Leshin, Lester L. Lyles, Paul D. Spudis, Neil deGrasse Tyson, Robert S. Walker, and Maria T. Zuber. 2004. *A Journey to Inspire, Innovate, and Discover: Report of the President's Commission on Implementation of United States Space Exploration Policy, 2021.* Available from http://www.nasa.gov/pdf/60736main_M2M_report_small.pdf; accessed March 18, 2005.

17. Alexander, Amir. 2008 (October 16). "President Signs NASA Authorization Bill." Planetary.org. Available from http://www.planetary.org/programs/projects/space_advocacy/20081016.html; accessed March 28, 2009.

18. Courtland, Rachel. 2009 (February 19). "Obama Backs Moon Return in NASA Budget." *New Scientist.* Available from http://www.newscientist.com/article/dn16676-obama-backs-moon-return-in-nasa-budget.html; accessed March 28, 2009.

19. David, Leonard. 2003 (March 4). "China Outlines its Lunar Ambitions." *Space.com,.* Available from http://www.space.com/missionlaunches/china_moon_030304.html; accessed January 30, 2006.

20. Aljazeera. 2006 (January 25). "Russia Plans to Mine the Moon." Available from http://english.aljazeera.net/NR/exeres/65F240F9-CC80-4677-93D6-EBCACE8E7A45.htm; accessed January 30, 2006.

21. Wakefield, Julie. 2000 (June 3). "Researchers and Space Enthusiasts See Helium-3 as the Perfect Fuel Source." *Space.com*. Available from http://www.space.com/scienceastronomy/helium3_000630.html; accessed January 30, 2006.

22. Wikipedia. 2009. "Nuclear Fusion." Available from http://en.wikipedia.org/wiki/Nuclear_fusion; accessed March 28, 2009

23. Robinson, George S., and Harold M. White, Jr. 1986. *Envoys of Mankind*. Washington, DC: Smithsonian Institution Press, 190.

24. Stormer, John A. 1964. *None Dare Call It Treason*. Florissant, MO: Liberty Bell Press.

25. United States Senate, "Moon Treaty," 70.

26. Reagan, Ronald. 1982 (June 8). "Speech to the House of Commons." Available from http://teachingamericanhistory.org/library/index.asp?document=449; accessed January 21, 2006.

27. United States Senate, "Moon Treaty," 122.

28. Ibid., 265.

29. United Nations. 1996. "Declaration on International Cooperation in the Exploration and Use of Outer Space for the Benefit and in the Interest of all States, Taking into Particular Account the Needs of Developing Countries." G.A. Res. 51/122. Available from http://www.unoosa.org/oosa/SpaceLaw/spben.html; accessed February 22, 2006; and Foreign Agriculture Service, United States Department of Agriculture. [n.d.]. "The General Agreement on Tariffs and Trade and the World Trade Organization (GATT/WTO)." Available from http://www.fas.usda.gov/itp/Policy/Gatt/gatt.html; accessed October 23, 2008.

30. O'Donnell, Declan J. 1999. "Property Rights and Space Resources Development." Available from http://www.mines.edu/research/srr/ODonnell.pdf; accessed February 1, 2005.

31. Ibid.

32. Ibid.

33. Fountain, Lynn M. 2003. "Creating Momentum in Space: Ending the Paralysis Produced by the "Common Heritage of Mankind" Doctrine." *Connecticut Law Review*, 35:1753. Available from http://0-web.lexis-nexis.com.opac.sfsu.edu/universe/document?_m=11fc26b13c17a600a2aaf5b31ce697fb&_docnum=28&wchp=dGLbVtb-zSkVb&_md5=2b3f6fb9a388d7b1b4b0f72e7eebeda1; accessed January 25, 2006.

7

Castles in the Air:
The Space Settlement Prize

EYES ON THE PRIZE IN THE SKIES

Alan Wasser, chairman of the Space Settlement Institute, proposes federal legislation requiring that the United States recognize extraterrestrial claims to real property, based on a unilateral reinterpretation of the Outer Space Treaty.[1] The Space Settlement Initiative appears on the National Space Society (NSS) Web site, although the NSS has not endorsed it. Both the Artemis Society and the Moon Society have endorsed the initiative.[2] Wasser's scheme for huge land claims on the Moon and Mars goes back a long way, as does his antipathy for the Outer Space Treaty and his public statements distorting the history of that treaty and international law in general. His idea of extraterrestrial land claims found its way into print as early as 1988.[3] By 1997, Wasser had drafted legislation.[4] By 2004, he had formed the Space Settlement Institute, with himself as chairman, his son David as executive director, and Douglas O. Jobes as president. The mission of the institute is to lobby for the passage of the Space Settlement Prize Act (SSPA).[5]

> It will take billions of dollars to develop safe, reliable, affordable transport between the Earth and the Moon. Neither Congress nor the taxpayers wants the government stuck with that expense. Private venture capital will support such expensive and risky research and development ONLY if success could mean a multi-billion dollar profit. Today, there is no profit potential in developing space transport, but we have the power to change that. We have the power to create a "pot of gold" on the Moon, waiting for whichever companies are the first to establish a "space line" and lunar settlement by risking their own necks, money and sweat.

How? By making it possible to claim and own —and re-sell to those back home on Earth—the product that has always rewarded those who paid for human expansion: *land ownership.*

The proposed legislation would commit the U.S. to granting that recognition if those who establish the settlements meet specified conditions, such as offering to sell passage on their ships to anyone willing to pay a fair price.

The first settlement on the Moon should be able to claim up to 600,000 square miles. Getting to Mars will cost much more and Mars itself is larger than the Moon. Therefore the first Martian settlement should be able to claim up to 3,600,000 square miles, roughly the size of the United States, worth 230 billion dollars at even $100 per acre.[6]

We should be trying to find a Congressional representative to introduce legislation saying that, while the U.S. makes no claim of national sovereignty, until and unless a new treaty on outer space property rights is adopted, all U.S. courts are to recognize and defend the validity of a land claim by any private company (or group of companies) which met the specified conditions.[7]

It is rare for Congress to take the initiative in foreign policy. Congress usually defers to the President. Only in cases where the administration's foreign policy seriously troubles domestic politics will Congress react strongly enough to affect the policy. For example, Congress attempted to reverse the Reagan administration's "constructive engagement" policy toward the apartheid regime in South Africa.[8] The issue of outer space property rights is hardly likely to rise to the level of public interest at which thousands of people take to the streets to demand a change in U.S. foreign policy. Thus, Congress will do nothing. If Wasser were able to persuade a member of Congress to introduce his bill, it would likely attract no cosponsors and would be referred to a subcommittee, never to be heard from again. In the unlikely event that Congress passed Wasser's bill, the President would summarily veto it. First, the President would veto it because all administrations adamantly defend the executive branch's historical prerogative in foreign policy; second, because no administration would acquiesce in national legislation contrary to longstanding U.S. foreign policy, since doing so would erode presidential authority over foreign policy; and finally, the State Department would vehemently oppose a bill that it regarded as being a treaty violation. By any calculation, "trying to find a Congressional representative to introduce" the SSPA is a fool's errand. This is simply not how the foreign policy apparatus of the United States works.

> Since it would not cost anything, or need any appropriations, such legislation might pass as a minor revision of property law.[9]

What Wasser proposes is not "a minor revision of property law"; it is a major foreign policy initiative that reverses 40 years of unwavering

U.S. commitment to the Outer Space Treaty. Since that treaty is the bedrock of international space law, the cost of unilateral national legislation aimed at diluting the treaty would be incalculable in terms of destabilizing the entire framework of international space law. It can be assumed that many states would be hostile to such a unilateral act, and rather than "adopt similar laws," states would be far more disposed to enact national legislation repudiating all private property claims in outer space. Forcing an issue usually polarizes the situation. Far from promoting commercial space development by removing a supposed barrier, very real barriers would be erected. If anything, commercial space activity would be likely to contract in this atmosphere of political hostility and legal uncertainty. Positions on this issue would harden, and it might take decades for them to soften to the point where meaningful negotiations could take place. Rather than a space Renaissance, Wasser's proposal would plunge space development into a Dark Age.

> One extreme would be a unilateral assertion of national authority. For example, the U.S. government could abrogate the Outer Space Treaty and declare sovereignty over some or all of the moon (while noting that the American flag is already there). Or, a bit more subtly, a government might extend recognition to certain property claims (probably those of its own citizens or companies) even without claiming sovereignty over the territories in question. Such unilateralism, however, would generate international tensions and, in all likelihood, competing claims by foreign governments. The rights of the supposed property owners—recognized by the courts of only one country—could hardly be said to exist.[10]

Oblivious to the danger, Wasser "would personally like to see" the United States withdraw from the Outer Space Treaty.[11] On other occasions, however, he alleges that his scheme does not violate the treaty because it does not require the United States to exercise sovereignty.

THE LAND CLAIM RECOGNITION ANALYSIS

Jobes and Wasser's "Land Claim Recognition (LCR) Analysis"[12] is replete with misunderstood and misapplied points of law. Furthermore, they appear to have little knowledge of the U.S. government or how its foreign policy apparatus operates:

> The appropriate legal framework for land claims recognition in space is the "use and occupation" standard from civil law. Use and occupation means the claimants, by establishing a permanent presence on the land, have

mixed their labor with the soil and created property rights that are independent of government.

In civil law countries like France, property rights have never been based on sovereignty as they have in the U.S. (which inherited the "common law" standard from the U.K.). Even in the U.S, derivatives of civil law are used by some states. From the *New American Encyclopedia:*

> Common law was generally adopted in the U.S., although Louisiana state law is based upon the Code Napoleon, and other states have partially codified systems. Civil law often relies on precedent, just as many common law rules are codified by statute [as in civil law] for convenience.
>
> Use and occupation must be the standard for any land claims regimen in space, because the common law standard cannot be applied on a Moon where sovereignty itself is barred by international treaty. Congress will have to decree that, because there can be no government on the Moon, a permanent base or settlement can give itself title just as though it were a government.[13]

There are a number of problems with the above passages. First, the United States is one of those countries whose legal system derives from common law, so how could it legitimately espouse a civil law theory of property rights in outer space? Although it is true that Louisiana, as a former territory of France, has a legal tradition that descends from civil law, Louisiana law is not federal law, and Jobes and Wasser aim at the extraterritorialization of federal law to the Moon or Mars. In this context, Louisiana law, whatever its tradition may be, is irrelevant. So, if "use and occupation must be the standard for any land claims regimen in space, because the common law standard cannot be applied on a Moon where sovereignty itself is barred by international treaty," this puts Jobes and Wasser between a Moon rock and a hard place. Moreover, Congress cannot "decree" anything. It may pass bills, which, if signed by the president, become law. As a common law nation, "because there can be no government on the Moon, [if a] permanent base or settlement [were to] give itself title just as though it were a government," it is hard to see how the United States could recognize any such title. The legal concept is incompatible with the legal system of the United States. On this basis alone, any U.S. court is likely to shoot down such legislation. Jobes and Wasser write as though the U.S. legal system were under the complete and direct control of Congress. Have they not heard of the "separation of powers" principle?

Second, the civil law concept that mixing labor with the soil creates property rights is inconsistent with Wasser's earlier suggestion that wealth could be created "out of thin vacuum."[14] Understandably, they would like to have their green cheesecake and eat it, too.

Finally, if "use and occupation means the claimants, by establishing a permanent presence on the land, have mixed their labor with the soil and created property rights that are independent of government," why is it necessary for any government to legislate in this matter? In the absence of government, the right exists by virtue of use and occupation, and the firepower to ensure the continuance of use and occupation. However, this implied use of force is a function of government. For "a permanent base or settlement [to] give itself title just as though it were a government," that base or settlement would have to *be* a government. What is a government? In the present system of nation-states, a government is what the governments of other nation-states say it is. The legitimacy of any government depends in large part on its recognition by other governments. Thus, ultimately, for Jobes and Wasser's ideas to have any specie in the nation-state system, the states of Earth would have to recognize lunar and Martian states. Such ideas may be vehicles for B-grade science fiction films, but they have little thrust as a basis for public policy.

> Participation by other space-faring nations would also help demonstrate these activities are in compliance with the "benefit of all mankind" requirement. Land claim recognition legislation could even direct the U.S. State Department to negotiate treaties requiring the private entities to form multinational consortia, to assure other nations that land claim recognition is not just an American attempt at a Lunar land grab.[15]

If the few launching states devour all of the outer space goodies, that is not "for the benefit of all mankind," it is for the benefit of *some favored few* of mankind. Furthermore, pursuant to Article II, Section 2, Clause 2, of the Constitution of the United States, the president

> shall have Power, by and with the Advice and Consent of the Senate, to make Treaties, provided two thirds of the Senators present concur.

The House of Representatives has no role at all in treaty making, and the Senate has only the reactive role of giving its advice and consent, not a proactive role. Both houses have standing committees on foreign relations, but their purview is limited to the general legislative power of oversight and investigation. Congress may direct the State Department all it wants, and the secretary of state, being answerable only to the president, may nod politely and go about regular business.

HISTORICAL ISSUES

Wasser has been playing fast and loose with history at least since 1988:

President Johnson was said to be in earnest for such a resolution because of the uncomfortable situation which existed since the Soviet Union soft-landed the first spacecraft on the lunar surface—*Luna 9* on 3 February 1966. Though the United States landed *Surveyor 1* four months later on 2 June; as there was no precedent for appropriation of extraterrestrial territory, it was unknown what approach the Soviets might take, and how international law would eventually rule.

Though this Treaty purports to endorse the development of the Moon and space in general, it contributes no fiscal or policy support to such developments. In fact, it places severe, probably prohibitive restraints, on such developments.

Wasser suggests that the opportunities and pace of space development would be greatly accelerated if the Articles were amended to say that, "Any nation, corporation or person that establishes a permanent manned settlement on the Moon or other celestial body, can claim all the land for 100 miles around (or 500, or 1000)."[16]

In 1997, Wasser again points to the landing of *Luna 9* as the triggering event for the Outer Space Treaty, and this time he also identifies the treaty as the triggering event for the collapse of the U.S. space program:

On February 3rd, 1966, the Soviet Union's *Luna 9* made mankind's first "soft" landing on the Moon. The U.S. was still trailing in the "space race" and wouldn't make its first soft landing for four more months.

To be able to divert more money to the escalating Viet Nam war, Lyndon Johnson had to send Arthur Goldberg to the Russians to negotiate a quick truce in the space race. The result was the 1967 Outer Space Treaty which, among other things, barred claims of "national sovereignty" in space.

The treaty doesn't actually bar private ownership of land beyond the Earth, but since national sovereignty has traditionally been the legal basis for private property rights in Anglo-Saxon law, the treaty is often assumed to have that effect.

I am convinced that that treaty provision is the real reason the space race ended, and space development has slowed to a crawl for the last quarter century. Significantly, space funding increased every year, in both the U.S. and U.S.S.R., until the passage of the 1967 treaty, and then decreased every year thereafter.[17]

It is easy to account for Wasser's conviction that "that treaty provision is the real reason the space race ended." He is no scientist. If he were, he would understand that correlation is not necessarily evidence of cause and effect. The universe is replete with unrelated coincidences, and any effort to relate two events without scientific rigor leads to myth and superstition: my neighbor's cow was struck by lightning, therefore he must have angered and been punished by the gods. The real reason for the end of the space race was that the United States so clearly won it. The space race was a venue of Cold War competition between the United States and the Soviet Union, an opportunity to show the rest of the world whose technology was better. Once the point was made, in terms of national interest, there was no point in continuing to make the same point. To put it in language that any 10-year-old can understand, the United States went to the Moon on a double-dare. And, unsurprisingly, the United States displayed an attention span comparable to a 10-year-old; we are a young country.

In saying, "No congress would ever have spent 100 Billion Dollars (in 1994 dollars) just for some nebulous 'prestige' benefit," Wasser only shows that he is unacquainted with Max Weber's three main dimensions of social life: wealth, power, and prestige.[18] These are things that individuals and nation-states alike covet. Why did the United States spend $500 billion (2005 dollars) and 58 thousand lives on Vietnam? It certainly was not for profit, so it must have been for power or prestige. For the United States to have or to not have power over what Johnson called "a raggedy-ass little fourth-rate country" was relatively inconsequential; neither the U.S. global strategic position, nor even the U.S. strategic position in eastern Asia, was substantially weakened by the fall of the Indochinese dominoes. What really suffered throughout the 1970s as a result of the loss in Vietnam was U.S. prestige. As it turned out, we could not "pay any price, bear any burden, meet any hardship, support any friend, oppose any foe, in order to assure the survival and the success of liberty,"[19] and bumping up against that bruising reality was a bitter blow to U.S. pride. On the other hand, landing the first humans on the Moon was a sorely needed source of U.S. prestige abroad, although it was a less-effective salve to the U.S. domestic psyche.

Wasser asserts that Johnson wanted to "negotiate a quick truce in the space race" to "divert more money to the escalating Viet Nam war." As can be seen in Figure 7.1, canceling the U.S. space program outright and giving all of NASA's budget to the Department of Defense would hardly have been noticed, either then or since. The historical fluctuations of defense spending dwarf the NASA budget.

Also, looking at only the funding history misses the programmatic details. The decline in the NASA budget starting in 1967 was due to the fact that the major capital investments in infrastructure (assembly facilities, test

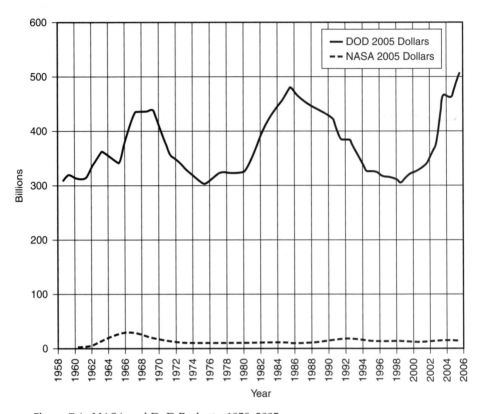

Figure 7.1: NASA and DoD Budgets, 1958–2005

Sources: sae.org and en.wikipedia.org/wiki/NASA_Budget.

chambers, transportation systems, launch pads, etc.) to support the Apollo program had been completed, as had most of the engineering design work for these ground support systems and flight systems. When one considers that from 1968 to 1972, the United States flew 11 Apollo missions at a time when the NASA budget (which of course included funding for programs other than Apollo) averaged $18 billion in 2005 dollars, compared to the peak budget in 1966 of $31.4 billion in 2005 dollars, it becomes obvious that the United States could have continued to explore the Moon indefinitely and relatively inexpensively.* That the United States did not do this had nothing at all to do with the Outer Space treaty,

---

* It should be noted that the 2005 NASA budget was $16 billion, or about 90 percent of its average budget during the Moon mission years; yet this is considered a starvation budget that barely keeps the Space Shuttle and International Space Station alive.

and very little to do with the Vietnam War. The programmatic decisions affecting the lunar exploration program, specifically the cancellation of the last three Apollo missions, came in 1970, three years after ratification of the Outer Space Treaty, during the time when the U.S. involvement in Vietnam was winding down. Harrison Schmitt (R-NM) refers to other programmatic decisions in 1972:

> In 1969 and 1970 there was a tremendous amount of other plans. We were working on all sorts of things.
> The momentum seemed to be for the United States to move forward with a continuous lunar exploration program and implicit in that is the potential exploitation of resources on the Moon in the foreseeable future, whereas in 1972 that all changed. The Nixon administration cancelled all those activities and it became obvious to the world that we would not do that.[20]

In short, the United States went to the great expense of building a railroad to the Moon that could have operated for many years and eventually could have supported profit-making enterprises, but it chose instead to tear up the tracks and scrap the rolling stock. This was a political decision, unrelated to treaties and wars. At best, Nixon was ambivalent about continuing the exploration of the Moon; to some extent, it served to bolster U.S. global prestige, but it also served to keep the Kennedy mystique alive.

One argument that Wasser uses to support his case against the current treaty regime is that in the 1960s the United States was concerned that the Soviet Union would land humans on the Moon first and would use this to establish a claim to the Moon; thus, in essence, the United States was allegedly frightened into the position of championing the concept of outer space as a commons and into signing a bad treaty. Supposedly, following the landing of the Soviet robotic spacecraft *Luna 9* on February 3, 1966:

> Newspapers ran serious articles about whether the Russians would use their landing to claim ownership of the Moon. Government officials worried about the supposedly overwhelming military advantage the U.S.S.R. would gain by seizing the "ultimate high ground." Reassuringly, the articles concluded that under traditional international law, no one could really claim the Moon until they had at least made a manned landing!
> Those articles are an excellent reminder that fear of a Russian victory in the race to the Moon, leading to a Russian claim to the Moon, was a major reason Congress kept increasing the funding for Apollo. No congress would ever have spent 100 Billion Dollars (in 1994 dollars) just for some nebulous "prestige" benefit.[21]

There is a great deal of distortion here. First, if some "newspapers ran serious articles about whether the Russians would use their landing to claim ownership of the Moon," they were seriously misinformed:

> Debate about whether to use air or high seas analogies ended in 1961, when the Soviet government accepted General Assembly Resolution 1721A. It expressed preliminary agreement on two key propositions:
>
> (a) International law, including the Charter of the United Nations, applies to outer space and celestial bodies.
> (b) Outer space and celestial bodies are free for exploration and use by all States in conformity with international law and are not subject to national appropriation.[22]

Furthermore, in its Draft Declaration of the Basic Principles Governing the Activities of States Pertaining to the Exploration and Use of Outer Space, which the Soviet Union submitted to COPUOS on September 10, 1962, it declared in Principle 2:

> Outer space and celestial bodies are free for exploration and use by all States; no State may claim sovereignty over outer space and celestial bodies.[23]

At the time, the Soviet Union had flown four manned missions in Earth orbit, the longest of which was 3 days, 22 hours, 22 minutes *(Vostok 3);* the United States had flown only two orbital missions, the longest of which was 4 hours, 56 minutes *(Mercury Atlas 7).* Soviet crews had accumulated 7 days in space; U.S. crews only 10 hours. The Soviet Union had impacted *Luna 2* on the lunar surface and had photographed the never-before-seen far side with *Luna 3.* By any quantitative or qualitative measure, the Soviet Union was winning the "space race." Yet, the Soviet Union had taken the position that "no State may claim sovereignty over outer space and celestial bodies."

So whatever some newspapers might have been publishing, the Johnson administration and Congress knew full well that the Soviet Union had committed itself to the freedom of outer space five years earlier. In fact, no extant principle of international law would have prevented the Soviet Union from claiming the Moon in 1959 when *Luna 2,* carrying a number of medallions with the national coat of arms, impacted the Moon. "The Soviet Union would have had the right to claim vast areas of outer space as its own territory, including part or even all of the Moon, based on historical precedents of exploration and conquest."[24] The Soviet Union made no sovereignty claims on the Moon with the *Luna 2* in 1959, nor with the *Luna 9* impact landing in 1966, nor was there any apprehension that it would do so if it were to land the first crew on the Moon.

Wasser is also misleading when he makes it sound as though Arthur Goldberg rushed off to Moscow to cut a hasty deal. Quite the contrary, the Outer Space Treaty was the culmination of years of groundwork, having as its progenitors UN General Assembly Resolution 1721 (XVI) in 1961 and Resolution 1962 (XVIII) in 1963 (the International Cooperation Resolution and the Declaration of Legal Principles, respectively). These resolutions, as well as less-prominent ones of the early Space Age, "were without exception the outcome of prolonged preliminary political negotiations and compromises between the two space powers."[25] It is also true that they were without exception adopted without vote, and that "[t]he operative parts of the resolutions 1721 and 1962 clearly show the crystallization on the legal principles" of outer space years before Ambassador Goldberg allegedly ran off "to the Russians to negotiate a quick truce in the space race."[26] Resolution 1962 provided "the international community with norms, in respect of which, the Member States of the United Nations, including the space Powers, have achieved consensus."[27] The correct reading of history is that the 1967 Outer Space Treaty was the extension of that consensus and the elaboration of those principles, not the result of some hasty, last-minute scramble.

Finally, rather than the United States having been anxious to conclude an early treaty:

> The Soviet Union, supported by Socialist Countries and at first by some other States too, was in favour of a treaty-declaration. The United States, on the other hand, wanted a General Assembly resolution (resolution-declaration) and not a treaty. The Soviet argument [was] partly based on the relevant provision of the UN Charter according to which resolutions of the General Assembly are without legal binding force, and partly on the suggestion, that the agreed principles of space law should be incorporated in a legally binding instrument (treaty).[28]

From this we see that, both counterintuitively and counter to Wasser's assertions, at the time of its greatest technological lead over the United States, the Soviet Union was anxious to establish an air-tight legal system for outer space, whereas the United States desired more flexibility. However, this was a difference over the form of the agreement, not over the principles themselves. The fact is that the United States desired outer space to be internationally recognized as a commons beginning in the early 1950s, long before it became apparent that the Soviet Union had an initial advantage in space launch capability (see Chapter 2, "Property Rights in Outer Space"). Thus, to say that the United States would have preferred a treaty that allowed the appropriation of space is an inaccurate reading of the historical record, and this false assertion cannot be used to advance the idea that the United States should reverse its position after 50 years, or to subvert, violate, or withdraw from the Outer Space Treaty.

Another historical point to be made is that John F. Kennedy's May 1961 commitment to "achieving the goal, before this decade is out, of landing a man on the Moon" was deliberately chosen as the most spectacular goal in human spaceflight that the United States could be most confident in accomplishing ahead of the Soviet Union. In other words, it was clearly beyond the near-term capabilities of either country, and the United States would have plenty of time to catch up to and surpass the Soviet Union . The United States did exactly that. At the midpoint of the Gemini program at the end of 1965, as work continued on the Outer Space Treaty, the United States held every important advantage over the Soviet Union on the road to a human landing on the Moon: heavy-lift launch capability (Saturn 1 and Saturn 1B), fuel cell technology to power long-duration missions (*Gemini 4*), mission duration sufficient to reach the Moon and return to Earth (8 days on *Gemini 5*, surpassed by 14 days on *Gemini 7*), rendezvous of two manned spacecraft (*Gemini 6A* and *Gemini 7*), and extravehicular activity (EVA, *Gemini 4*). In contrast, at that point the Soviet Union had only demonstrated battery power of insufficient duration for a lunar mission (5 days on *Vostok 5*), no rendezvous capability, and it never man-rated its Proton launch vehicle. When the Outer Space Treaty was opened for signing on January 27, 1967, the Gemini program had completed six manned rendezvous and docking missions with numerous EVAs, and the launch of the first Apollo mission was expected to occur within weeks.*

THEORETICAL ISSUES

Wasser's idea of bestowing nation-sized land grants to corporations is without theoretical grounding. Locke expressed the labor theory of property ownership:

> It is labour then which puts the greatest part of value upon land, without which it would scarcely be worth any thing: it is to that we owe the greatest part of all its useful products; for all that the straw, bran, bread, of that acre of wheat, is more worth than the product of an acre of as good land, which lies waste, is all the effect of labour: for it is not barely the plough-man's pains, the reaper's and thresher's toil, and the baker's sweat, is to be counted into the bread we eat; the labour of those who broke the oxen, who

---

* Ironically, the crew of *Apollo 1* died that evening when a fire swept through their spacecraft.

digged and wrought the iron and stones, who felled and framed the timber employed about the plough, mill, oven, or any other utensils, which are a vast number, requisite to this corn, from its being feed to be sown to its being made bread, must all be charged on the account of labour, and received as an effect of that: nature and the earth furnished only the almost worthless materials, as in themselves. It would be a strange catalogue of things, that industry provided and made use of, about every loaf of bread, before it came to our use, if we could trace them; iron, wood, leather, bark, timber, stone, bricks, coals, lime, cloth, dying drugs, pitch, tar, masts, ropes, and all the materials made use of in the ship, that brought any of the commodities made use of by any of the workmen, to any part of the work; all which it would be almost impossible, at least too long, to reckon up.[29]

By this theory, the Moon has no present value because no one is presently working the land. More important, however, is the principle by which common property becomes private property: the land must be worked.

But, what counts as sufficient labour? Extremes are clear (spitting in an ocean is insufficient to count as labour, for example, and fully cultivating an unowned piece of land is obviously sufficient). Is it sufficient to expend labour in the first place, or must the labour be of a certain type?

Having raised these doubts, though, we can see that one limit to property ownership appears to be the capacity to work the land, i.e., if I can't labour this land, I can't have it.

[There is also t]he spoilage limitation: I am not entitled to take land from the commons if it will spoil or go to waste. If the products of labour spoil, I have taken more than my share. I am required remember to improve the land; in allowing land to go to waste or to spoil, I am not improving the common stock.[30]

Quite simply, in pushing for continent-sized land grants, Wasser shows no understanding of these philosophical issues:

In countries like France, which follow what is called "civil law" (as opposed to "common law" which the U.S. inherited from the U.K.) property rights have never been based on territorial sovereignty but on the "Natural Law" theory that individuals mix their labor with the soil and create property rights independent of government, which merely recognizes those rights.[31]

If we accept the natural law concept that labor with the soil creates property rights, a small settlement on the Moon would create property rights for itself in its vicinity, rights to the land that is uses and occupies. Logically, this argument cannot extend property rights to claims the size of Alaska or the United States.

Some critics object that would allow a settlement to claim more land than it can use, but the amount of land that can be used depends on what you are using it for. Nineteenth century land grant farmers used 40 acres and a mule. Modern mechanized farms use vastly more land than that. Cattle ranchers use much more land than farmers. But none of those are the size criteria that should be used for a Lunar settlement because, of course, the settlement will not make its living by either farming or ranching.

The plan is to let settlements recoup the cost of getting there in the first place by selling land. If you are in the real estate business, especially if you are selling totally raw Lunar land, you can use all the acreage you can get title to.[32]

This reasoning is utterly preposterous. Speculating in a commodity does not equate to using a commodity for a productive purpose. If Wasser cornered the world market in pork bellies and beef, this would be less productive than cooking and serving a single bacon cheeseburger. The latter activity creates added value, whereas speculation is simply gaming the system to extract existing value from it, often to the detriment of someone else, as with Enron's gaming of the energy market to create artificial scarcities and price spikes in California in 2000–2001.[33]

No person (either natural or juridical), simply by virtue of establishing regular transportation between the Earth and the Moon, would instantaneously acquire the capacity to improve a surface area the size of Alaska; therefore, it could not immediately appropriate so vast a territory from the commons (even were there a sovereign to recognize such a title). Furthermore, there can be no presumption that the entity could acquire sufficient capacity to work the land before the period terminated during which it had the sole capacity to operate a translunar transportation system. This gives rise to the possibility that, were the entity to have title to this extensive area, this would deny use to others who later acquired the capacity to work claimed but still-unused land. This would constitute a taking of land from the commons and letting it go to waste. Obviously, the idea of a land grant on Mars the size of the United States also fails these tests. Jean-Jacques Rousseau reduces Wasser's position to absurdity:

> In general, to establish the right of the first occupier over a plot of ground, the following conditions are necessary: first, the land must not yet be inhabited; secondly, a man must occupy only the amount he needs for his subsistence; and, in the third place, possession must be taken, not by an empty ceremony, but by labour and cultivation, the only sign of proprietorship that should be respected by others, in default of a legal title.
>
> In granting the right of first occupancy to necessity and labour, are we not really stretching it as far as it can go? Is it possible to leave such a right

unlimited? Is it to be enough to set foot on a plot of common ground, in order to be able to call yourself at once the master of it? Is it to be enough that a man has the strength to expel others for a moment, in order to establish his right to prevent them from ever returning? How can a man or a people seize an immense territory and keep it from the rest of the world except by a punishable usurpation, since all others are being robbed, by such an act, of the place of habitation and the means of subsistence which nature gave them in common? When Nuñez Balboa, standing on the sea-shore, took possession of the South Seas and the whole of South America in the name of the crown of Castile, was that enough to dispossess all their actual inhabitants, and to shut out from them all the princes of the world? On such a showing, these ceremonies are idly multiplied, and the Catholic King need only take possession all at once, from his apartment, of the whole universe, merely making a subsequent reservation about what was already in the possession of other princes.[34]

Thus, although Wasser is not as audacious as Dennis Hope* in claiming the whole universe from his apartment, it is a short journey from one position to the other, both of which are constructed on the same philosophical quicksand.

LEGAL ISSUES

In a number of Wasser's statements, he appears confused about what he is advocating:

I believe that, for claiming ownership over space real estate, some form of the classic Roman legal principal called *pedis possessio* will have to apply. *Pedis possessio* means that the first entity to set foot (*pedis*) upon and occupy a space can claim possession, and most of the world's property rights laws are based on that principal.[35]

What Wasser apparently does not understand is that the *pedis possessio* principle applies to *res nullius*; one may take possession of that which belongs to no one by setting foot on it. However, the body of international law has clearly established that outer space and celestial bodies are *res communis*; as such, they are owned by the human community, and no one may take exclusive possession.

---

* Dennis Hope began selling "property" on the Moon in 1980 at $20 per acre, and claims to have had nearly 3.5 million customers, including two former U.S. presidents. He also claims every other celestial body in the solar system with the exception of the Earth and the sun.

Article 1 of the Outer Space Treaty provides that "there shall be free access to all areas of celestial bodies." How is "free access" compatible with Alaska-sized and even United States–sized land grants? Any meaningful property rights include the exclusive use of the owned land, negating "free access." Wayne N. White observes:

> [One] reason for prohibiting territorial sovereignty was to ensure free access to outer space. If nations begin claiming large areas of outer space or on celestial bodies, it will prevent entities from other nations from having free access to both claimed and unclaimed areas of outer space.[36]

Obviously, private appropriation on the scale of national territory would prevent free access as effectively as direct national appropriation.

Another problem with Wasser's concept surrounds the concept of "recognizing land claims":

> Under a land claim recognition protocol, Congress could pass legislation providing that for any private, non-government corporation or consortium that financed and built a space transportation system and permanent Moon base, a limited (but still very large) claim to lunar land around the base would be legally "recognized" by the U.S. government.
>
> Recognition means the government would acquiesce to, or decide not to contest, the claim, but not assume any sovereignty over it. Once the space transportation system and lunar base were certified, the private consortium would be free to immediately mortgage or sell, back here at home, some of their lunar land deeds to recoup their investment and make a profit.[37]

What would be the credibility of this so-called "recognition"? Would such a law obligate the United States to take action against those who did not "recognize" or otherwise violated a supposed property right? If so, such action would be an act of sovereignty. Before getting into the enforcement issue, let's tackle a more basic one. The fact that only states are parties to international agreements cannot be construed to mean that they have no bearing on nongovernmental entities. States bear international responsibility for the activities of nongovernmental entities under their jurisdiction. A state cannot license nongovernmental activities that are prohibited to the state. For example, the United States cannot get around the 1963 Test Ban Treaty by licensing a contractor such as Halliburton to detonate a nuclear device above ground. If states were to recognize a real property claim by a nongovernmental entity under its jurisdiction, this would constitute national appropriation by "other means," in violation of Article 2 of the Outer Space Treaty.

It has been suggested that states could unilaterally establish a domestic registry for the purpose of documenting the claims of their nationals to space resources, purportedly consistent with the non-appropriation principle. This "consistency" is provided by the artifice of proclaiming this registration scheme "not to be appropriation." For example, one group of proponents asserted that "[i]n doing so, the nation could make it clear that it was *not claiming* sovereignty over such resources, but *simply recognizing the claims* of its citizens (emphasis added)." This is a distinction without a difference.

Recognition of claims is only one side of the equation. The other side is the exclusion or rejection of any competing or conflicting claims. The application of this de facto exclusion of other states and their nationals by its very nature would constitute a form of national appropriation. Thus, state recognition of claims to extraterrestrial property by its nationals is national appropriation "by any other means" prohibited by article II, no matter what euphemistic label is employed to mask the obvious.[38]

With regard to enforcement, Wasser does intend to obligate acts of sovereignty:

The law could pledge to defend extraterrestrial properties by imposing sanctions against aggressors.[39]

It cannot be denied that imposing sanctions is an act of state. To defend extraterrestrial properties via this mechanism would therefore serve to support national appropriation by other means, in violation of Article 2 of the Outer Space Treaty. Furthermore, since such sanctions would have no legitimacy in international law, they would violate other international obligations. For example, suppose that a U.S. company established a permanent settlement on the Moon and claimed land according to the SSPA formula. Suppose that a foreign company later established a value-extractive operation one hundred kilometers from the U.S. settlement, within its land claim. This would in no way violate international law, yet would violate the SSPA, and could trigger the imposition of sanctions. The United States could impose the sanctions either against the foreign company directly, against the state in which the foreign company were based, or both. In any case, the foreign state, either on its own behalf or on behalf of the sanctioned company, would be well within its rights to bring a complaint against the United States before the World Trade Organization (WTO). There would be no shred of international law with which the United States could defend itself against such a complaint, and the WTO dispute settlement system would unquestionably find for the plaintiff. WTO member states are committed to having their national laws and regulations reviewed by the WTO. If the WTO rules against a state, it must abolish the law or pay a fine to the plaintiff country. If the state in violation does neither, the plaintiff country is empowered to enact trade sanctions against the offending country. In extreme cases, the WTO can encourage systemic sanctions against the offending country. U.S. companies that do international business are very aware of the power of the WTO, and it would be quite obvious to them that the threat of sanctions contained in the SSPA has no credibility. Thus, passage of the SSPA would not incentivize the commercial development of celestial bodies.

The major policy statements and legal provisions of the SSPA are contained in Sections 2 and 4. A number of paragraphs are merely declaratory, some of which are benign and are omitted from the following analysis, while others are contrary to international law. On the other hand, its legal provisions, in every instance, rest on the misinterpretation of existing law.

Section 2, paragraph 3, of the SSPA states:

> Space exploration and settlement with private financing will produce new tax revenues for the United States.[40]

This is almost certainly untrue in the short term. Among the arguments raised against the Moon Agreement was the prospect of an international regime that would levy such a heavy tax burden on commercial operations as to render them unprofitable. This argument is discussed and refuted in chapter five, "Death and Taxes." Since commercial operations involving human spaceflight are likely to remain only marginally profitable for quite some time once they reach profitability, it is extremely doubtful that Congress would discourage the development of a nascent industry by imposing tax liability. We see here how space property rights advocates talk out of both sides of their mouths; taxation by an international regime would squash the private development of space, but the private development of space would "produce new tax revenues for the United States." Well, let's get the story straight and stick to it—are private developers going to be able to pay their taxes or not?

Section 2, paragraph 4, of the SSPA states:

> A new, additional, incentive is needed because the potential short-term profit sources are currently much too small to attract the billions of dollars of private capital necessary.[41]

Here we have the answer to the question just posed. The profit margin will be so small that developers will not only be unable to pay taxes, they will need some sort of corporate welfare "incentive."

Section 2, paragraph 5, of the SSPA states:

> The potential value of land on the Moon, Mars, or an asteroid can provide an additional economic incentive for privately funded space settlement at no cost to the government.[42]

Ah, that magic word, "potential"! The Moon and Mars have no present value, since the means to go there, to work their lands, and to extract value from them, do not exist. At the present time, they are so much dust. But, of course, they have potential. In the "Financial Issues" section of this chapter, we explore Wasser's calculations of "potential value."

Section 2, paragraph 6, of the SSPA states:

> Prizes such as the Orteig Prize and the Ansari X Prize have an excellent record of promoting privately funded innovation, so Congress wishes to establish a "Space Settlement Prize" to promote the human settlement of the Moon and Mars.[43]

Burt Rutan's Scaled Composites spent $25 million on the *SpaceShipOne* project to win the $10 million Ansari X Prize, a loss of $15 million. Of course, Microsoft cofounder Paul Allen, whose estimated net worth is $21 billion, provided the venture capital, and what is $25 million to him?[44] Money was no object; the prestige of being first was the object. But Wasser does not believe in the allure of prestige.

Section 2, paragraph 7, of the SSPA states:

> At some time in the future Congress may be in a position to add an appropriately large monetary award, but, for now at least, the tremendous economic value of land claims recognition should be more than sufficient.[45]

Hold onto your wallets, Mr. and Ms. Taxpayer. Some day the space privateers will no longer be satisfied to joyride on Paul Allen's largess. That will be peanuts compared to the billions they can loot from you. And, they will have to because, as we see later, there is no "tremendous economic value of land claims recognition."

Section 2, paragraph 8, of the SSPA states:

> There is currently no international law on private land ownership in space, because most major nations have deliberately refused to ratify "The Agreement Governing the Activities of States on the Moon and Other Celestial Bodies, 1979, (hereafter called the "Moon Treaty"). The U.S. Senate's refusal to ratify means that the Moon Treaty's provisions are not "the law of the land" in U.S. courts, and therefore do not inhibit the actions of U.S. citizens or legislators.[46]

In fact, there is currently "international law on private land ownership in space." It is prohibited. Article 2 of the Outer Space Treaty states:

> Outer space, including the moon and other celestial bodies, is not subject to national appropriation by claim of sovereignty, by means of use or occupation, or by any other means.

As discussed in "Sovereignty and Property on Earth" in chapter two, a property right cannot exist in the absence of a controlling legal regime. There is no legal system outside of sovereignty except that which is established between sovereigns, that is, international law. In the absence of a legal system, there can be no legal title to anything. The Outer Space Treaty does recognize some forms of ownership. Article 8 provides:

> Ownership of objects launched into outer space, including objects landed or constructed on a celestial body, and of their component parts, is not affected by their presence in outer space or on a celestial body or by their return to the Earth.

How is such ownership possible if there is no sovereignty in outer space? In fact, there *is* sovereignty in outer space, not over territory, but over "space objects." Article 8 also provides:

> A State Party to the Treaty on whose registry an object launched into outer space is carried shall retain jurisdiction and control over such object, and over any personnel thereof, while in outer space or on a celestial body.

On the other hand, the Outer Space Treaty does not provide for the ownership of land. If it did so, it would first need to recognize the establishment of sovereignty over territory, which would be required to create the legal regime that would recognize the property right. But "national appropriation by claim of sovereignty" is expressly prohibited in Article 2, so the property right over territory cannot exist.

Section 2, paragraph 9, of the SSPA states:

> More importantly, the framers of the Moon Treaty found it necessary to attempt to write a rule forbidding private ownership of land on the Moon, clearly confirming that such an objective had not already been accomplished by "The Treaty on Principles Governing the Activities of States in the Exploration and Use of Outer Space, Including the Moon and Other Celestial Bodies", 1967, (hereafter known as the "Outer Space Treaty"), nor by U.N. resolution GA/res/1962.[47]

This statement is untrue on several points. First, as is made clear in the following statement before the Senate Subcommittee on Science,

Technology, and Space by Art Morrissey, senior policy analyst for the White House Office of Science and Technology Policy:

> The Moon Treaty is based to a considerable extent on the 1967 Outer Space Treaty. Indeed, the discussion in Outer Space Committee confirmed the understanding that the Moon Treaty in no way limits the provisions of 1967 Outer Space Treaty.[48]

Second, Wasser selects a single fact to support his erroneous conclusion and ignores the rest of the evidence. He fails to take note of the repetition and elaboration of principles not only from the Outer Space Treaty to the Moon Agreement, but from earlier sources, from the International Cooperation Resolution to The Declaration of Legal Principles to the Outer Space Treaty (see chapter two, "Legal Principles"). He asserts that a provision in a later document confirms that "such an objective had not already been accomplished" in an earlier document. It does no such thing. It is merely a restatement. Wasser misconstrues the purpose of repeating general provisions from one document to another, which is to provide continuity as well as to preclude fragmentation of the legal regime in cases where a state is party to one treaty and not another:

> To some extent, the trend toward fragmentation is limited by the fact that new space treaties generally repeat the general provisions which have already been endorsed by earlier treaties dealing with outer space. Although this legislative technique may raise difficult questions about the relationship between the obligations created by different instruments, it enables lawmakers to establish a legal system in which some basic rules are adopted by states which may not be bound by similar provisions in earlier treaties. As a result, the rules of space law acquire broader community support.[49]

Section 2, paragraph 10, of the SSPA states:

> The ratification failure of the Moon Treaty means there is no legal prohibition in force against private ownership of land on the Moon, Mars, etc., as long as the ownership is not derived from a claim of national appropriation or sovereignty (which is prohibited by the Outer Space Treaty).[50]

The failure of the Moon Agreement to be ratified by more than a handful of states, nonlaunching states at that, leaves the Outer Space Treaty as the source of the "legal prohibition in force against private ownership of land on the Moon, Mars, etc." This paragraph implies that a legal title of ownership could arise outside of sovereignty, but does not explain how.

Section 2, paragraph 11, of the SSPA states:

> Presumably it is only a matter of time until new treaties are negotiated,
> establishing a functional private property regime and granting suitable land
> ownership incentives for privately funded space settlements. The U.S. will,
> of course, abide by such new international law when it has ratified such a
> new treaty. But, given the urgent need for privately funded human expan-
> sion into space, as soon as possible, something must be done immediately,
> on a provisional basis, to correct the present inefficiencies in the interna-
> tional standard on property rights in space and to promote privately funded
> space exploration and settlement.[51]

In fact, functional property rights do exist under international law
(see chapter three, "Functional Property Rights"). The key word here is
"functional." For a property right to exist, something or someone must
be performing some value-extractive function on the land. The idea of
granting continent-sized land titles to corporations has no relevance to
this principle whatsoever. Since no entity has the capacity to perform
value-extractive functions on all of a continental landmass simultaneously, the theory of functional property rights cannot be used to advance
continent-sized land claims. This paragraph also asserts that there are
"present inefficiencies in the international standard on property rights in
space." What are they exactly?

Section 2, paragraph 12, of the SSPA states:

> For property rights on the Moon, Mars, etc., the U.S. will have to recognize
> natural law's "use and occupation" standard, rather than the common law
> standard of "gift of the sovereign," because sovereignty itself is barred by
> existing international treaty.[52]

Natural law is a legal theory, not a legal system. Theoretically, natural
law exists independent of recognition by a sovereign, so the proposition
that the United States "will have to recognize natural law" is double-talk.
Any law that the United States "recognizes" by act of Congress is by definition incorporated into the system of sovereign law, irrespective of its
origin in natural law theory. However, an act of Congress recognizing
property rights on the Moon or Mars would also be by definition an act of
sovereignty, "sovereignty itself is barred by existing international treaty."

Section 2, paragraph 13, of the SSPA states:

> U.S. courts already recognize, certify, and defend private ownership and
> sale of land which is not subject to U.S. national appropriation or sover-
> eignty, such as a U.S. citizen's ownership (and right to sell to another

U.S. citizen, both of whom are within the U.S.) a deed to land which is actually located in another nation. U.S. issuance of a document of recognition of a settlement's claim to land on the Moon, Mars, etc., can be done on a basis analogous to that situation.[53]

It is true that the U.S. legal system may have jurisdiction over certain cases in which the property in question is located in another nation, but the dispute over the property involves a U.S. citizen. However, the U.S. court finds its jurisdiction according to international law, not in violation of it. What is not in question in such cases is that there is a property that is subject to ownership by some entity under the sovereignty of some nation. On the other hand, the SSPA seeks to create a property right where none currently exists, in violation of international law; thus the SSPA's reference to these terrestrial cases is completely irrelevant.

Section 2, paragraph 14, of the SSPA states:

> This legislation concerns only the issuance of such a U.S. recognition and acceptance of a settlement's claim of *private* land ownership based on use and occupation, regardless of the nationality of the owner, and nothing in it is to be considered a claim of national appropriation of, nor sovereignty over, any outer space body, or any part thereof.[54]

As has been previously discussed, the assertion that the "settlement's claim of private land ownership [is] based on use and occupation" of millions of square kilometers is a fallacy. Wasser incorrectly equates buying and selling with use and occupation. Second, merely asserting that nothing in his proposed legislation "is to be considered a claim of national appropriation of, nor sovereignty over, any outer space body" does not make it true. If the character of an act depended only on its characterization by the actor, no crime could be prosecuted. It is analogous to asserting, "I will kill you, but nothing I do is to be considered an act of murder." Tell that one to the judge.

Section 2, paragraph 15, of the SSPA states:

> The U.S. does not claim the right to "confer" private land ownership, and the U.S. states it is most definitely not making any claim of "national appropriation by claim of sovereignty, by means of use or occupation, or any other means" as prohibited by the Outer Space Treaty.[55]

So, by this proposed legislation the United States makes no claims; it simply arrogates the right to commit acts while not claiming to commit them.

Section 4, paragraph 1, of the SSPA states:

> All U.S. courts and agencies shall immediately give recognition, certification, and full legal support to land ownership claims based on use and occupation, of up to the size specified in Sections 6.1, 6.2, and 6.3 below, for any private entity which has, in fact, established a permanently inhabited settlement on the Moon, Mars, or an asteroid, with regular transportation between the settlement and the Earth open to any paying passenger.[56]

To "give recognition, certification, and full legal support to land ownership claims" is an act of sovereignty for the purpose of appropriation, which is prohibited by Article 2 of the Outer Space Treaty.

Section 4, paragraph 2, of the SSPA states:

> For a land claim to receive such recognition and certification, the settlement must be permanently and continuously inhabited. The location and the population of the settlement may change, as long as there continues to be an inhabited settlement within the original claim.[57]

The conditions for recognizing the claim are irrelevant, since the claim itself has no basis in international law, and recognition under any condition is a violation of international law.

Section 4, paragraph 3, of the SSPA states:

> Deliberate abandonment of the settlement shall be grounds for invalidating land ownership recognition derived from that settlement, but there shall be no penalty for brief unintentional absences caused by accident, emergency, or aggression.[58] *How brief?*

This paragraph is superfluous, since any land ownership recognition is invalid in the first place.

Section 4, paragraph 4, of the SSPA states:

> Recognized ownership of land under this law shall include all rights normally associated with land ownership, including but not limited to the exclusive right to subdivide the property and sell portions to others, to mine any minerals or utilize any resources on or under the land, as long as it is done in a responsible manner which does not cause unreasonable harm to the environment or other people.[59]

The United States cannot guarantee rights where it has no legal jurisdiction. Where it exercises legal jurisdiction, it exercises sovereignty. To do so over lunar or Martian land area would be an act of sovereignty for the purpose of appropriation, which is prohibited by Article 2 of the Outer Space Treaty.

Section 4, paragraph 5, of the SSPA states:

> If the requirements of this law continue to be met, all rights, privileges, and responsibilities shall be immediately transferable by sale, lease, or other appropriate means to any other private entity.[60]

No rights can be transferable because no rights exist in the first place. Section 4, paragraph 6, of the SSPA states:

> As long as the required conditions continue to be met, U.S. recognition documents shall remain valid for 100 years or until the U.S. ratifies a treaty that establishes an international property rights regime which gives comparable reward to privately funded settlement, whichever comes sooner.[61]

Any U.S. recognition documents shall become invalid before the ink is dry.

Section 4, paragraph 7, of the SSPA states:

> The U.S. pledges to defend recognized extraterrestrial properties by imposing appropriate sanctions against aggressors, whether public or private. It pledges never to allow the sale to U.S. citizens of any extra terrestrial land which was seized by aggression. But it makes no pledge of military defense of recognized extraterrestrial properties.[62]

The United States has imposed appropriate sanctions against Cuba for half a century and has adhered to them with high fidelity, yet it is Fidel who holds the long-playing record. It is this paragraph in particular that exposes the SSPA for the farce that it is. Here on Earth, if someone runs you off your property, you call the law enforcement authorities. But on Mars, if someone runs you off your property, you call the U.S. Secretary of State in the hope that he or she will testify before the House Committee on International Relations and the Senate Committee on Foreign Relations in support of passing sanctions legislation of dubious effectiveness. Ultimately, a right must be supported by a credible enforcement mechanism if it is to have any meaning. Since the SSPA eschews "military defense of recognized extraterrestrial properties," the rights that it purports to create are a tissue of illusions.

FINANCIAL ISSUES

Let us for the moment put aside Wasser's distortion of history, his disinterest in legal theory, and his disregard for international law, and examine the economic rationale for his scheme.

So the "right" size for a claim is that size which is just large enough to justify the cost of developing reliable space transport and establishing a settlement, . . . but small enough to force the development of cost effective, affordable, transport, . . . and small enough to still leave room for future settlements.

That's how the proposed settlement sizes were derived. Real estate experts guessed at the minimum the land would bring when you could buy a ticket and get to it. Space experts guessed at what was the least that financially efficient private companies could hope to establish settlements for. The average settlement cost estimates, divided by the estimated average dollars per acre, gave the number of acres needed. Converted to square miles, that worked out to approximately 600,000 square miles on the Moon and 3,600,000 square miles on Mars.[63]

I am curious to know who these "real estate experts" and "space experts" are. On what assumptions did they base these guesses and how valid are these assumptions?

The Space Settlement Institute believes that the most valuable resource in space is Lunar and Martian real estate.[64]

My first reaction to this absurd claim was, "I wonder, have they taken a good look at Nevada lately?" Apparently missing my humorous intent, Wasser replied that Nevada turns out to be "an excellent analogy" and calculated a "conservative estimate" that at an average of $18,667.67 per acre, Nevada's 70,400,000 acres of real estate are worth:

$$\$18,667.67 \times 70,400,000 \text{ acres} = \$1,314,203,968,000$$

So, at the lowest price, that's One Trillion, three hundred million dollars.

The Space Settlement Institute is proposing a land claims recognition prize of 600,000 square miles on the Moon. That's as big as Alaska—and five times as big as Nevada—but still only 4% of the Lunar surface.

So, as you say, having taken a good look at Nevada, The Space Settlement Institute believes that the most valuable resource in space is Lunar and Martian real estate.[65]

Nevada is "an excellent analogy?" How so? Nevada may be in the middle of nowhere, but at least it has breathable air. The Moon and Mars are well beyond the outskirts of nowhere, and neither has breathable air. I and 40 million other Californians can get to Nevada in four hours on a tank of gasoline. I, or more likely, my ashes, can get to the Moon or Mars in 15 years on a $50 billion investment in developing flight hardware, software, and operations training. As for water and vegetation, Nevada is a veritable rain forest in comparison to the Moon and Mars. In terms of

Figure 7.2: Real Estate in Nevada and on Mars and the Moon

Source: elams.org and NASA.

proximity to major population centers and accessibility of resources, Nevada is orders of magnitude more valuable than the Moon or Mars (see Figure 7.2).

Another calculation Wasser ran was based on Dennis Hope's Lunar Embassy scheme:

> Since 1980 a man by the name of Dennis Hope has made a small fortune selling Lunar "deeds." He simply announced that he had claimed the Moon, set up his own "Lunar Embassy," and has sold unrecognized Lunar land "deeds" for $19.99 an acre ($22.49 if you want your name printed on the deed).
>
> Currently, Dennis Hope's website lunarembassy.com has sold over 2,300,000 Lunar "properties" to people in 165 countries. So Hope has proven

beyond doubt that real deeds, recognized by the U.S. and actually accessible by a then-existing commercial space line, would certainly be worth no less than $19.99/acre. 19.99 times 9,383,748,198 acres = $187,581,126,478. That is nearly $190 billion dollars—absolute minimum worst-case value.[66]

Wasser's reference of Hope hardly enhances the credibility of either of them. In any case, it takes more than simple arithmetic to understand the mathematics of economics. Wasser neglects basic economic principles, such as market size and price elasticities. The fact that Dennis Hope can take $20 each from thousands of people cannot be scaled linearly to infer the existence of the billions of buyers who would be required to finance Wasser's grandiose schemes. It is entirely invalid to extrapolate even a couple of orders of magnitude beyond the referenced data set. Also, a given person may buy an acre of lunar "property" and show the deed around to his friends as a novelty, but is he going to buy a thousand acres and thereby impress his friends with what an idiot he is? The per-acre price of a thousand-acre lot just isn't the same as the price of a one-acre lot.

My estimate of Wasser's business model is as follows: Dennis Hope claims that he has had "more than 3,470,072 customers" in the 26 years he has been in business. Let us stipulate that there are 3.5 million more untapped suckers in the world (or will be, according to Barnum's Law,* by the time the first privately financed lunar settlement is established). Let us also stipulate, for the moment, that Hope's going price of $20 per acre holds, despite the fact that this private entity, which has been cash-flow negative until this point and is desperate for revenue, has now glutted the lunar land market with 600,000 square miles of property for sale, rather than distributing the sales over a 26-year period. The company cannot afford to wait 26 years; it needs the money now! Since there are 640 acres in a square mile, this amounts to 384 million acres. This means that these 3.5 million potential buyers would have to buy an average of not just one acre, but 110 acres, for an average price of $2,200 per buyer.

The problem is that the market history is of 3.5 million customers over a 26-year period at a price of only $20. How credible is it that there will be a market of 3.5 million customers at a price of $2,200 over a period of, let us say, a year or two? Not very credible. Prices will be elastic, since no one on Earth needs to buy land on the Moon; this is an optional purchase. There will be substantially fewer than 3.5 million buyers, and prices will collapse.

---

* "There's a sucker born every minute."

So, let us come up with a more credible model, and speculate that there might be 350,000 people who would be willing to spend $220 on something that almost none of them will ever be able to see or touch, raising a grand total of $76 million. That might buy a secondhand space suit for someone who got to be an astronaut when he or she grew up.*

There is more to consider on the subject of the value of scarcity. Regarding his continent-sized land grants, Wasser points out:

> Fortunately, that is quite small enough to still leave plenty of room for subsequent settlements, since it is only around 4 percent of the Moon, 6.5 percent of Mars.[67]

How true. The surface area of the Moon is equal to all of South America, and the surface area of Mars is equal to all of the land area of Earth. This is hardly what one would call a scarce resource. So, what tangible difference is there between the unimproved land inside Wasser's property fence and the unimproved land outside it? I am reminded of the scene in Monty Python's *Life of Brian,* in which an entrepreneur sells rocks to the righteous along a rock-strewn road leading to the stoning of a blasphemer.[68]

> Of course, the establishment of their space transport service, which enabled the consortium to win the land grant in the first place, will dramatically increase the value of their land over what it is worth today, when it is inaccessible. As with the land grants that paid for building America's transcontinental railroads, vast wealth would be created (out of thin vacuum, so to speak) by giving formerly worthless land real value and an owner.[69]

There are several inaccuracies in this paragraph. First, wealth is never created "out of thin vacuum." Wealth is created from productive activity involving land, capital, and labor. In contrast, Wasser uses language that conjures visions of Ponzi schemes, where money from later investors is used to pay off earlier investors, but all such schemes ultimately collapse. The early investors make out like bandits because they have robbed the later investors, who end up with nothing.[70]

Also, it very plainly would have been impossible for "land grants [to have] paid for building America's transcontinental railroads." If the land over which the railroads were about to be built was worthless, it could not have been a source of capital for building the railroads. Wasser has confused cause and effect; the land began to acquire some value once the

---

* A new extravehicular suit costs between $300 million and $700 million.[71] Oxygen sold separately.

infrastructure was in place, once value had been added to the land by the productive application of labor and capital. Given the level of technology at the time, it obviously took a tremendous amount of human labor to build the transcontinental railroads; it also took a great deal of capital.

> In addition to the grant of lands and right of way, Government agreed to issue its thirty year six per cent. Bonds in aid of the work, graduated as follows: For the plains portion of the road, $16,000 per mile; for the next most difficult portion, $32,000 per mile; for the mountainous portion, $48,000 per mile.
> The Union Pacific Railroad Co. built 525 78/100 miles, for which they received $16,000 per mile; 363 602/1000 miles at $32,000 per mile; 150 miles at $48,000 per mile, making a total of $27,236,512.
> The Central Pacific Railroad Co. built 7 18/100 miles at $16,000 per mile; 580 32/100 miles at $32,000 per mile; 150 miles at $48,000 per mile, making a total of $25,885,120.
> The total subsidies for both roads amount to $53,121,632. Government also guaranteed the interest on the Companies' first mortgage bonds to an equal amount.[72]

In 2005 dollars, the $53,121,632 in 1865 dollars would be more than a billion dollars to build a railroad that private investors, not the taxpayers, own. Far greater subsidies and loan guarantees will be necessary to establish regular transportation service to and from a settlement on the Moon or Mars. These projects cannot possibly be financed with grants of currently valueless land, any more than the transcontinental railroads were this way. These unimproved (indeed, presently unimprovable due to their inaccessibility) land holdings will secure no present loans, will purchase no present material, and will pay no present wages, whatever their "guessed" future value may be. This would be true even if recognition were given to the land claim on the day that the project began, rather than after the success of the project had been proven. It is important to remember that, pursuant to Section 4, paragraph 1, of the SSPA, U.S. courts would only "give recognition, certification, and full legal support to land ownership claims based on use and occupation" once a private entity has "established a permanently inhabited settlement on the Moon, Mars, or an asteroid, with regular transportation between the settlement."[73] Unless and until these conditions were fulfilled, the private entity would own nothing at all.

> Once the space transportation system and lunar base were certified, the private consortium would be free to immediately mortgage or sell, back here at home, some of their lunar land deeds to recoup their investment and make a profit.[74]

On the other hand, perhaps the righteous would pick up one of the many rocks on the side of the road that are free for the taking; the condemned blasphemer cannot not tell the impact of a free rock from a purchased one.

The Space Settlement Initiative is so flawed in its historical analysis and justification, in its theoretical and philosophical underpinnings, in its legal grounding, and in its financial strategy, that the practical effect is to solicit money and/or effort from people for a political project that is exceedingly unlikely to succeed, and perhaps to provide a source of income to officers of the Space Settlement Institute.

NOTES

1. Wasser, Alan B. 2004. "The Space Settlement Initiative." Space Settlement Institute. Available from http://www.spacesettlement.org/; accessed March 19, 2005.

2. *Space Daily.* 2003 (April 30). "Moon Society and Artemis Society Endorse Space Settlement Initiative." Available from http://www.spacedaily.com/news/mars-base-03b.html; accessed March 17, 2005.

3. Wasser, Alan B. 1988 (February 5). "National Space Society Airs '67 Outer Space Treaty Reservations." *Space Daily.* Available from http://www.space-settlement-institute.org/Articles/archive/TreatyReservations.pdf; accessed March 25, 2005.

4. Wasser, Alan. 1997 (March). "How to Restart a Space Race to the Moon and Mars." *Moon Miners' Manifesto, 103.* Available from http://www.asi.org/adb/06/09/03/02/103/space-race.html; accessed March 19, 2005.

5. Wasser, 2004, "Space Settlement Initiative."

6. Wasser, Alan. 2001 (October 9). "The Space Settlement Initiative." *Space Future.* Available from http://www.spacefuture.com/archive/the_space_settlement_initiative.shtml; accessed December 26, 2005.

7. Wasser, "How to Restart a Space Race."

8. Treverton, Gregory F., and Pamela Varle. 1992. "The United States and South Africa: The 1985 Sanctions Debate." Institute for the Study of Diplomacy, School of Foreign Service, Georgetown University.

9. Wasser, "How to Restart a Space Race."

10. Silber, Kenneth. 1998 (November). "A Little Bit of Heaven: Space-Based Commercial Property Development." *Reason.* Available from http://www.findarticles.com/p/articles/mi_m1568/is_n6_v30/ai_21231184; accessed March 19, 2005.

11. Wasser, "How to Restart a Space Race."

12. Jobes, Douglas O., and Alan B. Wasser. 2004. "Land Claim Recognition (LCR) Analysis: Leveraging the Inherent Value of Lunar Land for Billions in Private Sector

Investment." Space Settlement Institute. Available from http://www.space-settlement-institute.org/Articles/LCRbrieftext.htm; accessed March 16, 2005.

13. Ibid.

14. Wasser, "How to Restart a Space Race."

15. Jobes and Wasser, "Land Claim Recognition (LCR)."

16. *Space Daily.* 1988 (February). "National Space Society Airs '67 Outer Space Treaty Reservations." Available from http://www.space-settlement-institute.org/Articles/research_library/TreatyReservations.pdf; accessed March 24, 2005.

17. Wasser, "How to Restart a Space Race."

18. Weber, Max. 1947. *The Theory of Social and Economic Organization.* Translated by A M Henderson and Talcott Parsons. New York: Oxford University Press.

19. Kennedy, John F. 1961 (January 20). "Inaugural Address."Available from http://www.jfklibrary.org/j012061.htm; accessed January 3, 2006.

20. United States Senate. 1980. "The Moon Treaty." Hearings Before the Subcommittee on Science, Technology, and Space of the Committee on Commerce, Science, and Transportation. 96th Congress, 2nd Session, 54.

21. Wasser, "How to Restart a Space Race."

22. Peterson, M. J. 1997. "The Use of Analogies in Developing Outer Space Law." *International Organization,* 51, 2: 245–74.

23. United Nations Committee on the Peaceful Uses of Outer Space. 1962. U.N. Doc. A/AC.105/L.2.

24. Tennen, Leslie I. 2003. "Commentary on Emerging System of Property Rights in Outer Space." United Nations–Republic of Korea Workshop on Space Law. Available from http://www.oosa.unvienna.org/SAP/act2003/repkorea/presentations/specialist/ost2/tennen.doc; accessed March 19, 2005.

25. Csabafi, Imre. 1965. "The UN General Assembly Resolutions on Outer Space as Sources of Space Law." *Proceedings, 8th Colloquium on the Law of Outer Space.* American Institute of Aeronautics and Astronautics, 337–361.

26. Ibid; and Wasser, "How to Restart a Space Race."

27. Csabafi, "UN General Assembly Resolutions."

28. Ibid.

29. Locke, John 1689. *The True Original, Extent, and End of Civil-Government,* 43. Available from http://www.constitution.org/jl/2ndtr00.htm; accessed December 26, 2005.

30. University of Bristol Department of Philosophy. 2005. "John Locke (1632–1704): Property rights and their limits?" Available from http://www.bris.ac.uk/philosophy/current/undergrad/unitdesc/yr1/0506/soccontracttheory/lecture4.html; accessed December 26, 2005.

31. Wasser, 2001, "Space Settlement Initiative."

32. Ibid.

33. Isaacs, Jerry. 2002 (May 10). "Enron Defrauded California out of Billions During Energy Crisis." World Socialist Web Site. Available from http://www.wsws.org/articles/2002/may2002/enro-m10.shtml; accessed October 24, 2008.

34. Rousseau, Jean-Jacques. 1762 *The Social Contract*, I 9. Translated by G. D. H. Cole. Available from http://www.constitution.org/jjr/socon.htm; accessed December 28, 2005.

35. Wasser, Alan. 2004 (October 18). "The Space Settlement Prize." Space Review. Available from http://www.thespacereview.com/article/249/1; accessed December 26, 2005.

36. White, Wayne N. 1998. "Real Property Rights in Outer Space." *Proceedings, 40th Colloquium on the Law of Outer Space*, 370. American Institute of Aeronautics and Astronautics. Available from http://www.spacefuture.com/archive/real_property_rights_in_outer_space.shtml; accessed March 19, 2005.

37. Wasser, "Space Settlement Prize."

38. Tennen, Leslie I. 2003. "Commentary on Emerging System of Property Rights in Outer Space." United Nations–Republic of Korea Workshop on Space Law. Available from http://www.oosa.unvienna.org/SAP/act2003/repkorea/presentations/specialist/ost2/tennen.doc; accessed March 19, 2005.

39. Wasser, "How to Restart a Space Race."

40. Wasser, Alan. 2004. "The Space Settlement Prize Act." Available from http://www.spacesettlement.org/law/; accessed January 28, 2005.

41. Ibid.

42. Ibid.

43. Ibid.

44. Wired.com. 2004. "What Rich Guys Buy." Available from http://www.wired.com/wired/archive/12.07/start.html?pg=9; accessed February 18, 2006.

45. Wasser, "Space Settlement Prize Act."

46. Ibid.

47. Ibid.

48. United States Senate, "Moon Treaty," 29.

49. Danilenko, Gennady M. 1989. "Outer Space and the Multilateral Treaty-Making Process." *Berkeley Technology Law Journal*, 4, 2. Available from http://www.law.berkeley.edu/journals/btlj/articles/vol4/Danilenko/HTML/text.html; accessed March 19, 2005.

50. Wasser, "Space Settlement Prize Act."

51. Ibid.

52. Ibid.

53. Ibid.

54. Ibid.

55. Ibid.

56. Ibid.

57. Ibid.

58. Ibid.

59. Ibid.

60. Ibid.

61. Ibid.

62. Ibid.

63. Wasser, 2001, "Space Settlement Initiative."

64. Wasser, David, e-mail message to author, January 26, 2005.

65. Wasser, Alan, e-mail message to author, March 5, 2005.

66. Ibid.

67. Wasser, 2001, "Space Settlement Initiative."

68. Chapman, Graham, John Cleese, Terry Gilliam, Eric Idle, Terry Jones, and Michael Palin. 1979. *Life of Brian.* DVD and VHS. Produced by Tarak Ben Ammar. Directed by Terry Jones. Orion Pictures Corporation.

69. Wasser, "How to Restart a Space Race."

70. United States Securities and Exchange Commission. 2001. "'Ponzi' Schemes." Available from http://www.sec.gov/answers/ponzi.htm; accessed February 18, 2006; and United States Securities and Exchange Commission. 2004. "Pyramid Schemes." Available from http://www.sec.gov/answers/pyramid.htm; accessed February 18, 2006.

71. Harris, Garry L. 2001. *The Origins and Technology of the Advanced Extravehicular Space Suit.* American Astronautical Society History Series, Vol. 24. San Diego: Univelt, Inc.

72. Crofutt, George A. 1871. "Crofutt's Trans-Continental Tourist's Guide." Available from http://libr.unl.edu:2000/westward_through_nebraska/CG1871w.3.html; accessed February 18, 2006.

73. Wasser, "Space Settlement Prize Act."

74. Wasser, 2004, "Space Settlement Prize."

8

Celestial Empire: China's Rise as a Space Power

SATISFIED POWER

The rise and fall of the American space program can be understood in terms of its transition from the bipolar era of the Cold War, in which it was one of two superpowers, to "the unipolar moment," in which it is perhaps the global hegemon. The European age of exploration in the sixteenth and seventeenth centuries was a struggle for survival between contending powers, none of which were preeminent. Likewise, the early Space Age was a contest between the United States and the Soviet Union. Once the Americans beat the Soviets in landing humans on the Moon, and it became clear that the Soviet Union would never follow, the political raison d'être for the human exploration of space disappeared. The Apollo program ended without a follow-on program to continue the exploration of the Moon.

The Space Shuttle program emerged as a bureaucratic compromise between a NASA that had been bloated by Apollo and was frantically scrambling for some next grand development project, a Department of Defense that had been itching for a piece of the human spaceflight action for more than a decade, and various other federal agencies that had space ambitions. The budgetary reality of the 1970s denied NASA the money to fund such a large development project merely to keep humans flying in space. A major rationale of the Space Shuttle was as a transportation system to deploy and retrieve national security space assets, to extend the U.S. Empire into near-Earth space, to seize the heights in order to command the ground below. This was its black* justification. However,

---

* Sensitive compartmented information (SCI), the most highly-classified national security programs; as opposed to non-SCI (white) programs.

when OV-099 *Challenger* blew up in 1986, the national security community decided to pull the plug on its involvement in the Space Shuttle program. A Space Shuttle launch complex at Vandenberg Air Force Base, meant to launch reconnaissance satellites into near-polar, sun-synchronous orbits* from which they could surveil the entire planet, was mothballed without ever having been used.** Plans for enlarging the Space Shuttle fleet with "blue" (i.e., Air Force–owned) orbiters evaporated. National security satellites that had been specifically designed for the Space Shuttle, and that could not be reconfigured for other launch systems, flew on the first few missions when the Space Shuttle fleet returned to flight in 1988. Since then, what started out as a substantially black program in a white wrapper has probably been mostly white, ironically emerging out of tragedy to become what the U.S. taxpayer thought it had been all along.

The white justification for developing the Space Shuttle fleet was that it would serve to ferry crews and supplies to a space station. From a programmatic standpoint, the transportation system *and its destination* should have been developed simultaneously. Without a space station, such a large, manned vehicle made little sense, at least in the white world, and the fact that such a space station did not even begin to be assembled until nearly 20 years after the Space Shuttle became operational points up the primacy of the Space Shuttle's black justification during its development.

In another historical irony, the Space Shuttle system failed just as its primary white reason was needed most. As a result of OV-102 *Columbia*'s breakup in 2003, the partially-built International Space Station (ISS) limps along in orbit, tended by a skeleton crew who can do little more than maintain it, much less perform the science duties that justify its existence. Although it will probably be "completed" someday, according to some definition, the ISS is not the exploration of space, it is merely the

---

* A sun-synchronous orbit is one that combines semimajor axis and inclination so that a spacecraft passes over any given point of Earth's surface at the same local solar time. The surface illumination angle will be nearly the same at every passage over that point. This consistent lighting is useful for satellites that image Earth's surface in visible or infrared wavelengths, such as meteorological, reconnaissance, and remote sensing satellites.
** This was the second time the Space Launch Complex 6 was mothballed. It was originally built to launch the Manned Orbiting Laboratory (KH-10), an Air Force reconnaissance platform based on NASA's Gemini spacecraft, which was canceled in 1969 in favor of the less-expensive, unmanned, film-returning spacecraft often referred to in open sources as Big Bird (KH-9). In the early 1980s, SLC-6 was reconfigured for Space Shuttle launches carrying, among other things, digital-imaging KH-11s. In the early 1990s, SLC-6 was reconfigured yet again, this time for Athena launch vehicles. Following only four Athena launches, SLC-6 is now being reconfigured to launch Delta IVs, making it perhaps the most expensive and least efficiently used real estate in history

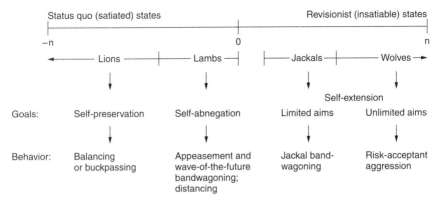

Figure 8.1: Schweller's Linear Schema

Source: Schweller 1994.

occupation of space, and moreover, a region of space that we have come to know quite well. It boldly goes where hundreds of people have been going for nearly half a century.

This is not to say that the United States has engaged in no space exploration since the Apollo program, but simply that it has engaged in no human space exploration. Even its robotic exploration of the Solar System has been anemic, and most of these accomplishments were due to the momentum of Kennedy's New Frontier. In the first 20 years of the Space Age, the United States sent missions to explore the Moon and every planet in the Solar System. Since the Apollo missions to the Moon, human planetary exploration has not been necessary as a demonstration of national technical prowess. The United States has not needed to outdo anyone. Even the Soviet copy of the Space Shuttle, the Buran program, was canceled in 1993 before the first manned flight.* In space, at least, the United States is a satisfied nation-state, the consummate "lion" in Randall Schweller's zoological description of nation-states (see Figure 8.1):

> Lions are states that will pay high costs to protect what they possess but only a small price to increase what they value. . . . As extremely satisfied states, they are likely to be status quo powers of the first rank.[1]

Unsurprisingly, what we have seen from the United States since the development of the Space Shuttle in the 1970s is status quo behavior in

---

* The only flight (unmanned) of the program occurred on November 15, 1988.

space. Moreover, the scrapping of the Saturn V launch vehicle and the NERVA upper stage has been likened to the decommissioning of the Chinese Navy in the fifteenth century. The Ming government had the capability to launch an age of exploration nearly a century ahead of Europe, yet stepped back from this adventure. China was a lion, a satisfied power. In the centuries that followed, it declined and was overrun by foreign powers.

ENTER THE DRAGON

Do not count on China making the mistake of missing out on the next great age of discovery. In the mid-twentieth century, China recovered its independence and has been a rising power ever since. At the turn of the twenty-first century it is hardly a satisfied power. It is poised to become the East Asian hegemon and has irredentist claims on Taiwan as well as other territorial claims. In Schweller's menagerie, China may be a jackal or a wolf, but certainly not a lamb (see Figure 8.2):

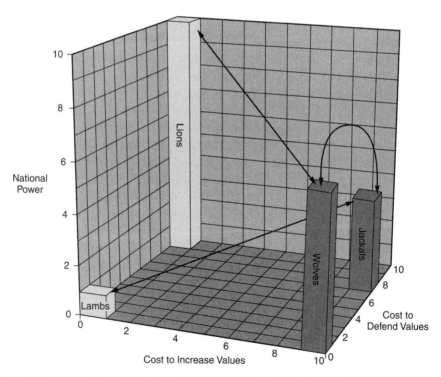

Figure 8.2: Three-Dimensional Augmentation of Schweller's Schema

Source: Gangale, Thomas. 2003. "Alliance Theory: Balancing, Bandwagoning, and Détente." San Francisco State University. Presented at the annual convention of the International Studies Association, San Diego, California, March 25, 2004.

Jackals are states that will pay high costs to defend their possessions but even greater costs to extend their values.

Wolves are predatory states. They value what they covet far more than what they possess.

Lambs are countries that will pay only low costs to defend or extend their values. . . . Lambs are weak states.[2]

Or, perhaps, China defies Schweller's zoological categories and is a unique animal—the dragon? U.S. proponents of scuttling or circumventing the international legal regime of outer space appear to assume that it will be U.S. private interests that will be quick to take advantage and to establish the first space settlements. However, the American Century is behind us. China became a space-launching state in 1970, barely a decade after the Soviet Union and the United States. For several decades its space activities were constrained by its underdeveloped economy, however, estimates are that China's gross domestic product (GDP) will surpass that of the United States in 30 to 40 years.[3] Meanwhile, its developing manned space program has caused barely a stir in the United States. Reportage of its first mission, *Shenzhou 5*, piloted by Yáng Lìwei on October 16, 2003, shrugged it off as repeating something the Soviet Union and the United States had accomplished four decades earlier. This is far off the mark. The Shenzhou design is not based on mid–twentieth-century technology, but on technology from the turn of the twenty-first century. It appears to be more versatile than the Soyuz spacecraft that the Soviet Union/Russia has flown for four decades, and it remains to be seen how the Orion, with which NASA *Cancelled* plans to replace the Space Shuttle, will measure up. The orbital module at the front of the Shenzhou is reportedly capable of being connected to others as building blocks for a small orbital station or as add-on modules to a larger station.[4] China is also reported to be interested in lunar and Mars exploration.[5]

The economic trends of the United States and China have a large bearing on the question of how long the present "unipolar moment" of the United States will endure. Estimates are in the range of 15 to 30 years.[6]

Table 8.1: Human Spaceflight Entities

Launching Entity	First Mission		
	Spacecraft Name	Crew	Launch Date
USSR/Russia	*Vostok 1*	Yurij Gagarin	April 12, 1961
USA	*Mercury Redstone 3*	Alan Shepard	May 5, 1961
China	*Shenzhou 5*	Yáng Lìwei	October 15, 2003
Scaled Composites	*SpaceShipOne*	Michael Melvill	June 21, 2004

Michael Mastanduno, professor of government at Dartmouth University, speculates that the unipolarity that has existed since 1990 has the potential to last at least this long for two reasons. First, the technological primacy, military and economic power, and ideological appeal combine to offer the United States strong potential to remain the world's only superpower. Second, unipolarity is the best position in anarchy compared to the uncertainty of a multipolar world and the concentrated hostility of a bipolar world. The unipolar state can preserve its dominant position by reassuring, engaging, and accommodating the other states and thus convincing them that the existing international order is desirable and acceptable.

Stephen Walt, academic dean of the Kennedy School of Government at Harvard University, argues that once a power achieves an overwhelming superiority, lesser states cease their attempts to balance.[7] Folding this idea into Schweller's menagerie, which also includes wolves and lambs, I call this the house cat theory of international relations. In the Lockean state of anarchy, cats are solitary, territorial, and aggressive. Yet, under the protection of a human sovereign, they live under the same roof in relative peace. In essence, the sole superpower approaches the position of global sovereign, and the international order resembles less the anarchic state that realist theorists assume.

According to Mastanduno, the United States is following a consistent strategy to preserve its preponderance by strengthening its security commitments with, and military presence in, Japan and Germany; expanding NATO; engaging Russia and China into the global capitalist economy; and using multilateral mechanisms to gain international support for its objectives. The future durability of this international order depends on the ability of the United States to meet three challenges:

- To continue to discourage the rise of states that combine formidable economic and military capability with global ambition.
- To manage and minimize the arrogance of power—the preaching of virtues and imposition of values—which creates resentment.
- To maintain domestic support for the political and economic policies needed to preserve preponderance (military intervention in distant places, free trade that causes loss of U.S. jobs, giving America's closest allies a free ride in military operations and economic relations, and doing business with human rights violators like China).

The problem is that sustaining these policies can be very expensive, especially the distant military adventures and the free-riding allies. Taken too far, they can result in what Paul Kennedy of the London School of Economics calls "imperial overstretch," draining the hegemon's resources, and exciting the world's disgust over the arrogance of power.[8] There are those who would point to the U.S. involvement in Iraq as a case in point.

There are signs that China chafes at a unipolar world order headed by the United States:

> With the victory in the U.S.-started war against Iraq, there has emerged in the United States the advocacy of the establishment of new order of a unipolar world, so as to ensure "world peace under American domination". At the beginning, this view began to spread only among the academic circles, but recently, it has been openly promoted by certain American and British politicians, this cannot but arouse people to take it seriously. . . .
>
> The emergence of unipolar theory and new empire theory has exposed the astonishing paradoxical phenomenon in U.S. politics: Domestically, the United States stands for power restraint and takes pride in it. But in international relations, the United States opposes any restraints and despises the United Nations, violates international law, disrupts the democratic principle, runs counter to world people's anti-war wishes, pursues force as supremacy, and seeks freedom of unilateral action, some people even openly advocate empire domination over the world.[9]

Is a return to bipolarity a credible scenario for the twenty-first century, this time with China replacing the Soviet Union as U.S. adversary? Historically, China was content to expand to its natural borders. Its primary concern was security, not military expansionism (which in part explains the dismantling of the Ming fleet). China has been a nuclear power for 40 years, and it still has only a very small nuclear capability, sufficient to deter an attack, but far from threatening a first-strike capability against either the United States or Russia. The Chinese view Taiwan as an internal matter; they used to have it, and they just want it back. Beyond this, the Chinese Communist Party, having cut itself loose from its ideological moorings, has staked its future legitimacy on its ability to deliver continued economic development. It may be that China has learned the lesson from the twentieth century that in the long run, being aggressive does not pay, that what pays is making friends and playing well with others. Germany and Japan did not do this in the early twentieth century and paid dearly; since then they have learned to fit into the world system and have done very well. The Soviet Union did not try to fit in, and it is gone. Thus China's long-term policy may be integration, not only into the global economy, but into international political institutions as well, to fully embed itself in the structures of stability. Why would a nation of well over a billion people want to go to a lot of bother to militarily dominate a few million more? Alastair Iain Johnston of Harvard University writes that China, having been relatively aloof from international institutions in the 1970s and 1980s, while rising as an economic power in the 1990s and the first decade of the twenty-first century, has joined international institutions rapidly.[10] Taken together with other evidence, Johnston concludes that China does not overtly seek to challenge the post–World War II international order that the United States and its allies have constructed, but to prosper within it.

INTERPLANETARY INSTITUTIONS

G. John Ikenberry of Princeton University argues that the unipolar moment persists, despite realist theory prediction that the world would return to mulitipolar rivalry and balance against the dominant power, because of the character of American hegemony.[11] Without a common external threat, great powers should have returned to competitive strategies driven by the underlying structure of anarchy. Germany and Japan should have rearmed and broken their ties with the United States. NATO should have fallen apart, leading to a multipolar scramble for power. Instead, the 1990s brought a reaffirmed U.S.-Japan alliance in the wake of no common threat. Europe reduced rather than increased defense spending. Russia and China voice some opposition to the U.S. hegemonic global presence, but also seek to integrate themselves into the Western-oriented world system. This stable order is rooted in the relationship between the United States and the outside world. What Ikenberry calls the "American System" today is built on two grand bargains with other countries around the world:

- *The Realist Bargain:* The United States provides Asian and European states with security and access to its markets in exchange for these states being stable and supportive partners.
- *The Liberal Bargain:* Asian and European States accept U.S. leadership and the world political-economic system. In return, the United States limits its own power by binding itself to its partners (i.e., the U.S. makes its power safe to the world and in return the world agrees to live in the U.S. system).

Since World War II, the U.S. hegemonic order has become more stable because the rules and institutions have become more firmly embedded in the wider structures of politics and society. The durability of U.S. liberal hegemony rests on the constitution-like character of the institutions and practices of the order, which serve to reduce the "returns to power," which in turn lowers the risks of participation by strong and weak states alike. Institutions also exhibit an "increasing returns" character, which makes it more and more difficult for would-be orders and would-be hegemonic powers to compete against and replace the existing order and leader.

Why would a newly hegemonic state want to restrict itself by agreeing to limits on the use of hegemonic power? If the hegemonic state calculates that its overwhelming postwar power advantages are only momentary, an institutionalized order might "lock in" favorable arrangements that continue beyond the zenith of its power. Additionally, it can reduce the enforcement costs for maintaining order. The "lock in" effect derives from the phenomenon of increasing returns, the large costs of creating new institutions, the ability to learn from the institutions in solving problems, and the relations and commitments that institutions tend to create among nations. This is the calculation that the United States made at the end of World War II.

Because it is a liberal democracy, and because it operates in a world that is trending toward increasing numbers of liberal democracies, a world more integrated by international institutions, transnational organizations, global communication, and trade, in most cases the United States need not exercise *hard power*, and, indeed, it has many incentives not to. This process of systemic integration on a global scale necessarily means that the world is no longer as anarchic as realists have traditionally viewed it. Thus, pluralist and institutionalist theorists are closer to the mark in regarding the preeminent position of the United States in the world as characterized more by the exercise of *soft power*.[12] Because the world as a whole has yet to arrive at Francis Fukuyama's "End of History," realism may from time to time inform our dealings with nonliberal regimes who are still stuck in History.[13] But, if there is, indeed, an "End of History" and if we are traveling the path toward it, as we go down that road, the realist view becomes a vista that is receding into history.

In the meantime, we would do well to watch China. It is not clear whether China, were it given the opportunity to remake the world order, would or even could choose solutions similar to those made by the United States at the end of World War II. The United States was able to get agreement among the Western states on a mutually acceptable order by using its own example of a successful liberal democratic polity. The U.S. system invites the participation of outsiders. Can this be said of China? Keeping the existing institutions in place limits a future hegemon's ability to change the system to its sole advantage and to the detriment of the many.

The probable economic and political trajectories of the United States and China in the twenty-first century ought to inform our approach to shaping the future legal regime of outer space. To overturn the international legal regime of outer space, on the assumption that U.S. capitalism will be unleashed to grab the lion's share of extraterrestrial resources, is to bet the farm, and it is not a very safe bet. Given the degree to which the

United States has become a debtor nation due to decades of current account deficits, how valid can that assumption be? The U.S. trade deficit with China grew to $202 billion in 2005, with no sign of substantially decreasing anytime soon.[14] The United States is bleeding money, and China is soaking it up. In the long term, it may be that U.S. profligacy largely finances Chinese expansion into the solar system. Given that space technology for the economic development of the Moon and other celestial bodies is unlikely to be available before mid-century, by which time China will probably have surpassed the United States in GDP and may well rival the United States as a space power, it seems that opponents of the treaty regime of outer space would have us tear down the system just as it is needed most.

Although China has abandoned communism and embraced capitalism, it remains a repressive, single-party state, and as it has not embraced liberal democracy, it may not have run out of History just yet. It may indeed be committed to embedding itself in international political institutions at present, but this is uncertain at best, and in any case it is no guarantee of China's future policy. This points up the folly of throwing away the rulebook when one is too arrogant to follow it. In the long view, is it in U.S. interests, indeed, in the interests of industrialized, liberal democratic states generally, to abandon the legal principles of outer space? If we were to do so, by what international standards of behavior should we expect China to abide as it embarks on its exploration and exploitation of the Solar System? If the Outer Space Treaty were to become a dead letter, the way would be clear to conflicting claims not only of property, but of sovereignty. In case China were disposed to abrogate institutional arrangements and realign the international order to its own advantage through aggressive behavior, nothing could better encourage such future bad behavior than for the United States to provide the example that a superpower need not live by the rules, even the rules that it has helped to write for the world over the past half-century. Clearly, the wisest policy is not to tear down the international institutions of outer space, but to further embed China in them, and to work in partnership with China and other nations to further define these "rules of the spaceways" in a manner that creates greater legal certainty and encourages economic development.

NOTES

1. Schweller, Randall L. 1994. "Bandwagoning for Profit: Bringing the Revisionist State Back In." *International Security*, 19, 1:72–107.

2. Ibid.

3. Pesek, William Jr. 2005 (October 13). "China Is Neither Japan Nor the Soviet Union. *Bloomberg.com*. Available from http://www.bloomberg.com/apps/news?pid=10000039&sid=asQZkFCRv55I&refer=columnist_pesek; accessed March 16, 2006.

4. Jones, Morris. 2006 (February 19). "China Might Be Planning Early Space Station Attempt." *Space Daily*. Available from http://www.spacedaily.com/reports/China_Might_Be_Planning_Early_Space_Station_Attempt.html; accessed March 16, 2005.

5. McDonald, Joe. 2003 (October 16). "China Plans More Missions, Space Station." *Space.com*. Available from http://www.space.com/missionlaunches/china_plans_031016.html; accessed March 16, 2006.

6. Krauthammer, Charles. 1990. "The Unipolar Moment." *Foreign Affairs,* 70:23–33; Nye, Joseph S. 1990. *Bound to Lead: The Challenging Nature of American Power.* New York: Basic Books; Kapstein, Ethan B., and Michael Mastanduno. 1999. *Unipolar Politics: Realism and State Strategies After the Cold War.* New York: Columbia University Press; and Mastanduno, Michael. 1999. "A Realist View: Three Images of the Coming International Order." In *International Order and the Future of World Politics,* ed. T. V. Paul and John A. Hall. Cambridge: Cambridge University Press, 19–40.

7. Walt, Stephen. 2000. "Keeping the World 'Off Balance:' Self-Restraint and U.S. Foreign Policy." Kennedy School of Government Working Paper Number:RWP00-013. Available from http://ksgnotes1.harvard.edu/Research/wpaper.nsf/rwp/RWP00-013/$File/rwp00_013_walt.pdf; accessed September 30, 2003.

8. Kennedy, Paul. 1989. *The Rise and Fall of the Great Powers.* New York: Vintage.

9. *People's Daily Online.* 2003 (July 30). "The Falsehood of Monopolar Theory: Commentary." Available from http://english.people.com.cn/200307/30/eng20030730_121258.shtml; accessed February 23, 2005.

10. Johnston, Alastair Iain. 2003. "Is China a status quo power?" *International Security,* 27, 4. Available from http://bcsia.ksg.harvard.edu/BCSIA_content/documents/Johnston_spring_2003.pdf; accessed September 3, 2003.

11. Ikenberry, G. John. 1999. "Institutions, Strategic Restraint, and the Persistence of American Postwar Order," *International Security,* 23, 3: 43–78; Ikenberry, G. John. 1999. "Liberal Hegemony and the Future of American Postwar Order." In Paul and Hall, 123–145; and Ikenberry, G. John. 2001. "American Grand Strategy in the Age of Terror." *Survival,* 43, 4:19–34. International Institute of Strategic Studies.

12. Nye, Joseph S. 2004. *Soft Power: The Means to Success in World Politics.* New York: Public Affairs Group.

13. Fukuyama, Francis. 1992. *The End of History and the Last Man.* New York: Avon Books.

14. Scott, Robert E. 2006 (February 10). "Rapid Growth in Oil Prices, Chinese Imports Pump Up Trade Deficit to New Record." Economic Policy Institute. Available from http://www.epinet.org/content.cfm/webfeatures_econindicators_tradepict20060210; accessed March 16, 2006.

9

Interplanetary Political Economy

SPACE COWBOYS: BIG HAT, NO CATTLE

A number of space enthusiasts tout free enterprise as the wave of the future in space development, and take great delight in disparaging government space projects as building the wrong capabilities for too much money. I have no doubt that free enterprise will be important to space development, but I have grave doubts that it will have a significant impact in the near future. Yes, government programs are expensive, *Pre-SpaceX* because there are extraordinary engineering challenges to getting into space on the cheap.

It is one thing to accelerate to 3,500 km/hr and poke one's head above an arbitrarily defined threshold of 100 kilometers for a few seconds, as *SpaceShipOne* did in 2004. At the top of that steep parabola, horizontal velocity was near zero. It is orders of magnitude more difficult to reach half again as much altitude and simultaneously accelerate to a horizontal velocity of 29,000 km/hr. Even more important is the requirement to withstand the deceleration and heat-loading of reentry from orbital velocity. A rudimentary calculation will serve to demonstrate. The ratio of orbital velocity to the peak velocity of *SpaceShipOne* is 8.3 to 1; however, the energy is a function of the square of the velocity, thus the energy ratio is 68.7 to 1. Clearly, the total heat-loading experienced by *SpaceShipOne* is inconsiderable compared to return from orbit. Withstanding more heat requires more shielding, more shielding increases the mass of the payload (spacecraft), increasing the mass of the payload requires more propellant, increasing propellant means larger rockets, and all of the increases mean higher operational costs. Solving these problems in the context of a credible business model is decades away.

Some have defended the libertarian vision of private space development by mentioning such things as the federal Homestead Act. The *SpaceShipOne* flights are supposed evidence that private space tourism is not far in the future, and that private space travel to the Moon or Mars is not hopelessly romantic. The use of *in situ* resources, inflatable habitats, nanotechnology, and advances in computer technology and robotics should bring unanticipated capabilities and cost reductions.

And so might pixie dust. Engineering solutions are based on technology in hand, not unobtainium beyond the horizon. The question is: When? Some space enthusiasts point to the rapid improvements in computer technology—and the huge commercial industry it has spawned—as a model for projecting a coming explosion in commercial space travel. Rubbish! Nanotechnology is nothing more than the extension of Moore's Law from the micrometer realm into the nanometer realm. It has been anticipated for decades. It does not represent a sudden, steep upswing in the rate of technological progress. There was no sudden upswing during the computer revolution in the late 1980s, early 1990s; it was steady, incremental progress. Moore's Law of doubling chip capacity every two years has held for 40 years.

Now, here is the really bad news: there is no aerospace analogue to Moore's Law. There is no evidence that the cost per mass to orbit reduces by half over a specified time scale. In fact, the cost per mass to orbit remained virtually unchanged from 1994 to 2000 inclusive, a seven-year period. A September 2002 report by Futron Corporation concludes:

> Regarding [non-geostationary orbit] launch prices, there is no clear trend in the price-per-pound metric, other than a clustering around $10,000 per pound in the late 1990s. While this is lower than GSO launches, it is not as low as one might expect.[1]

At best, a linear trend might be inferred from the 1990–2000 data for geostationary orbit (GSO). Regressing the 1990 average of $40,740/kg and the 2000 average of $25,804/kg yields an average cost reduction rate of $1,494/kg/yr. The cost per mass to low Earth orbit (LEO) is less; however, the cost of operating a human-rated system is much higher because of the requirements for life support systems, higher reliability and safety standards, and most importantly, return capability. But let's be very charitable and assume that the cost of per mass of a human-rated system to LEO is the same as the cost per mass of a non–human-rated system to GSO (I doubt this is true, but consider it a government subsidy from the Gangale Republic). The cheapest human-rated Earth-to-orbit transportation system around (indeed, the only one operating at the moment) is the Soyuz 7K-STMA, which has a mass of 7,220 kg and can carry one paying passenger. At the 2000 average of $25,804/kg, a Soyuz flight should cost $186 million. Actually the going price for a Soyuz joyride has been about $20 million. Obviously, the commercial passenger is not bearing the full cost of the flight; he is just buying a seat and is just along for the ride, whereas the crew has a real mission to perform. This is not the paradigm of a commercial, profit-making venture.

But, let's take a flight of fancy and say that the Gangale government is going to throw in another subsidy and match the Russians' price of $20 million per seat. That is still sky-high (no pun intended). How is the stars-in-his-eyes space entrepreneur going to get the average person to subject himself to having his eyeballs shoved against the back of his skull and then have his stomach float up to his chest (not to mention the contents thereof floating about the cabin) for less than $2,000?* Bringing the spouse/domestic partner along doubles the price; bringing the kids just about doubles it again.

So, we need to reduce the seat cost from $20 million to $2,000, a difference of $19,998,000. At an average cost reduction rate of $1,494/kg/yr, that will take 13,400 years. But, as a final concession, let's throw in some unobtainium from the Gangale Republic's strategic reserve, and reduce this time frame by an unbelievable 99 percent. That will bring space tourism within reach of the average consumer in 134 years.

Here is another back-of-the-envelope calculation, much shorter, and with much less substantiation. Starting with the estimated full cost of a Soyuz flight, $186 million, and assuming for the sake of wild optimism the emergence of an aerospace analogue to Moore's Law with a time constant of 5 years (again, there is absolutely no historical basis for this), the $2,000 seat would become available about the year 2090.

Wayne N. White chooses to "respectfully disagree," insisting that:

> [I]t is wrong to extrapolate from the government dominated space programs of the past, making projections that don't allow for either advances in technology or the power of private competition. Whether its nanotech, computers, robotics or whatever, there will be tech advances.
>
> All this being said, we in the space activist community have been consistently disappointed that the pace of space development has been far slower than we have hoped or expected. As you can see, however, I remain idealistic and optimistic about the future. I think removing national and bureaucratic politics from the equation, and the benefits of competition, will result in much faster development in the future, particularly if we enact laws that enable rather than hinder private space activities.[2]

---

* I have some hours in fighter aircraft, and there is nothing comfortable about this kind of flight regime.

My analysis very clearly allows for advances in technology, based on historical data, and much of that technological advancement has been achieved through the power of private competition. That is how scientists work; whether it is natural science or social science, one makes inferences from the data. We do not engage in wishful thinking, then look for someone to blame when dreams don't come true on an unrealistic schedule. I took a degree in aerospace engineering because of my idealism and optimism regarding humankind's future in outer space; given that degree and the industry in which I worked, I make bold to say that I am at least as heavily invested as White is in an idealistic and optimistic vision of the future. I remain idealistic and optimistic about the future, which he may not see because I also endeavor to be realistic.

Here is a reality check from an independent source. In a September 2006 discussion with former Apollo astronaut and former L-5 Society president Philip K. Chapman, he expressed the hope that the cost per pound to orbit could be reduced to $1,000 in the next 20 years, possibly even as low as $500. No one wants to see space commercialization succeed more than Chapman. Before we try to translate this into the price of a seat, let us look at the historical data for orbital spacecraft mass versus maximum occupant capacity (see Table 9.1).

Early manned spacecraft had little to no capability for the significant orbital maneuvers that would be required for a commercial passenger

Table 9.1: Manned Spacecraft

Spacecraft	First Manned Flight	Launching State	Mass (kg)	Maximum Occupants	Mass per Occupant (kg)
Vostok/Voskhod	1961	USSR	5,682	3	1,894
Mercury	1962	USA	1,118	1	1,118
Gemini	1965	USA	3,851	2	1,926
X-20A Dyna-Soar	1966[a]	USA	3,600	1	3,600
Soyuz	1967	USSR	7,250	3	2,417
Apollo	1968	USA	5,806	3	1,935
Transportnij Korabl Snabzheniya	1981[b]	USSR	17,510	3	5,837
Space Shuttle	1981	USA	99,117	7	14,160
Buran	1994[c]	USSR	42,000	10	4,200
Shenzhou	2003	China	7,840	3	2,613
Orion	2014[d]	USA	13,886	6	2,314

[a] Schedule at time of program cancellation in December 1963
[b] Schedule at time of program cancellation in January 1978
[c] Schedule at time of program cancellation in June 1993
[d] Schedule as of October 2008

vehicle taking people to an orbiting space hotel. They also had few of the amenities that a commercial passenger would expect for a comfortable flight. Even so, let us take the lowest spacecraft mass per occupant (Mercury, at 1,118 kg) and Chapman's more optimistic figure of $500 per pound, and without even considering a cost factor for a man-rated system (additional safety requirements, life support systems, etc.), the calculation yields:

$$1{,}118 \text{ kg} \times 2.2 \text{ lb/kg} \times \$500/\text{lb} = \$1{,}229{,}800$$

This is the best-case, low-ball, no-frills fare—20 years from now. A more reasonable estimate would double the mass per occupant and double the cost per unit mass, yielding $5 million per seat.

White responds:

> Your opinion may be based on facts, but it is still an opinion, just like mine. We just reach different conclusions.[3]

That is an intellectually bankrupt position. The qualitative difference between his opinion and mine is that mine is based on fact, his is not. Furthermore, mine is based on not only historical data, but also engineering principles and experience, his is not.

One of the largest factors in determining the cost per mass to orbit is the mass fraction of payload to total launch mass. This can be improved somewhat by efficient structural design to reduce tankage mass, turbopump design, engine pressure, and some other things. However, these gains are small compared to the theoretical constrains on reaction mass exit velocity, which are a function of chemical bond energy. Reaction mass exit velocity is expressed as specific impulse, Isp. NASA's Glenn Research Center has a Web page that does a nice job of explaining it:[4]

> The higher the Isp, the more energy the fuel/oxidizer combination has, and the greater the mass fraction. The theoretical Isp limit for a hydrogen/oxygen bipropellant is around 460 sec., which is quite high on the scale of chemical Isp. Solid monopropellants are usually around 200–250 sec.; however, the fact that solids are cheaper to manufacture and easier to handle offsets their poorer performance. A higher Isp is a lithium-fluorine-hydrogen tripropellant (540 sec.), but fluorine is so corrosive that no one wants to mess with it; it would be so horribly expensive in the operational environment as to more than offset performance gains. Also, no one wants to use nuclear engines (Isp = 850 sec.) in Earth's environment, and some don't want to use them at all, so at best they would be restricted to use as transorbital stages. Ion engines are

very high Isp (3,000 sec.), but very low thrust devices, good for motoring around the solar system, but no good at all for getting on or off a planet.

When I was in engineering school at the University of Southern California 30 years ago, I read a fascinating article, either in *Aviation Week and Space Technology* or the AIAA's *Astronautics and Aeronautics*, about metallic hydrogen, in which theoretical Isp's of 1,000 to 1,600 sec. were discussed. I thought I'd take the conservative figure of 1,000 sec. and design a lunar tug around it. My professor said exactly two words before he turned away: "Totally unrealistic." He was right. Thirty years later, I haven't heard of metallic hydrogen as a propellant since, and we're still stuck using the same Isp's we had 40 years ago. The fundamental laws of chemistry have not changed in 40 years. We have been operating near the theoretical limits all that time, and there is no way past them. There is always a way to make a smaller circuit, but it is impossible to get much more energy out of a chemical reaction. That is why there is no astronautical analogue to Moore's Law.

So, these are the challenges that face the space tourism industry in getting into Earth orbit on the cheap. The Moon miners have a higher hurdle to get over, since the velocity to escape Earth's gravity entirely is twice orbital velocity. More velocity requires more propellant. Then, upon reaching the Moon, one must fire the engines once again to settle into orbit or land—more propellant.

Space is an extreme environment, in which humans and their engineering systems operate at the margins. The best analogue for spaceflight is not the personal computer, nor even the private aircraft, but the submarine. Submarines have been around for more than a century, yet there is no commercial submarine travel industry. *Untapped potential*

SOCIAL BALANCE IN SPACE

The golden decade of the 1960s has long since faded, and has been followed by more than three decades of disappointment and disillusionment in the space enthusiast community. If this community expected the United States to continue to devote 4 percent of the federal budget to the civil space program (versus the 1 percent average since), the expectation was unrealistic (see Figures 9.1 and 9.2.)* Not to have realized the

---

* Presidential candidate Rep. Dennis Kucinich of Ohio proposed tripling the NASA budget.[5] He garnered only 3.8 percent of the primary vote nationwide and received only 17 of the total 2,719 delegates to the 2004 Democratic National Convention.[6]

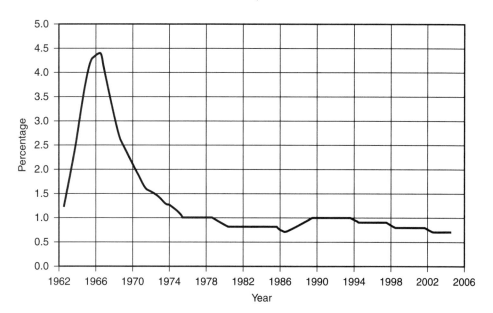

Figure 9.1: NASA Budget as Percentage of Federal Budget, 1962–2004

Source: www.asi.org/images/2003/NASA-budget-as-percentage-1962–2004.png

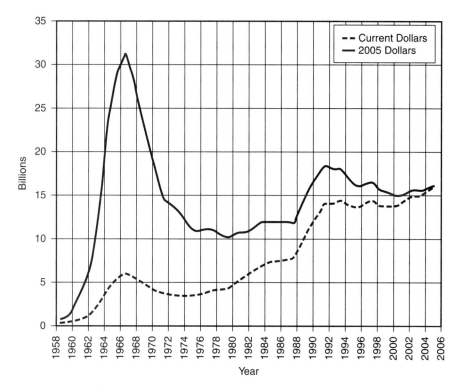

Figure 9.2: NASA Budget, 1958–2005

Source: en.wikipedia.org/wiki/NASA_Budget

limitations of technocracy at the time was naive, but understandable; failure to understand this several decades later is denial, which is inexcusable. To promote privatization as the panacea is to misunderstand the relative strengths and weaknesses of technocracy and technoeconomy; to advocate the destruction of the treaty regime to advance private interests is irresponsible. Responsible space advocacy requires an understanding of the political-economic forces that have brought us to where we are today, so that we can develop a coordinated strategy to harness these forces to best advantage to propel us toward the kind of future we want.

Technocracy, which is the state-directed "force-growing" of specific technologies to serve state interests, can achieve spectacular results in a short period of time (Figure 9.3). The problem with technocracy is its narrow focus—in time and in purpose. The purpose is power, not wealth. The time span is usually a few years—lasting only as long as it takes to achieve state interests, and the project is terminated once it no longer serves those interests. There may be residual economic benefits from the project, but these are incidental and are not the justification for the resources invested. The calculus is political, not economic.

Technoeconomy is the "organically grown" technologies developed for the market. These also serve a narrow interest—producing profit for the investor—but that purpose is long-lived. The project lasts as long as there is a market for the product line. As the science fiction author

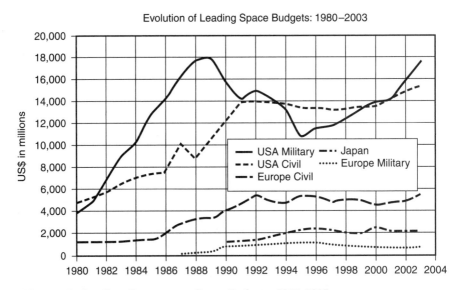

Figure 9.3: Leading Government Space Budgets, 1980–2005

Source: www.spacedaily.com/news/satellite-biz-03zzzl.html

Poul Anderson had his space entrepreneur Nicholas van Rijn say, "Politics, they come and go, but greed goes on forever."[7]

Technocracy and technoeconomy each have their strengths and weaknesses, and to a large degree they complement each other. It is a fundamental mistake to compare the accomplishments of technocracy and technoeconomy. "If we can put a man on the Moon, why can't we build a better mouse trap?" This is not a valid question, any more than asking, "If birds can fly, why can't fish climb trees?"

The Manhattan project built the first nuclear reactor in 1942. Twenty years later, power companies were telling the public that nuclear power plants would make electricity so cheap that it would not be worth metering, and everyone would pay a flat monthly fee. Today— more than 60 years later—the U.S. commercial nuclear power industry is struggling.

> In pricing electrical power produced by these plants, no one had factored in the extremely high hidden costs of disposing of dangerous radioactive materials and of finally sealing off the power plants themselves.[8]

The Vostok project put the first human in Earth orbit in 1961. Within five years, Stanley Kubrick was filming a motion picture that depicted Pan Am flights to a rotating Earth orbital space station and to the Moon. Today—more than 40 years later—no commercial effort has duplicated Yurij Gagarin's flight, nor is any likely to in the next decade. The flights of *SpaceShipOne* in 2004 reached the lower reaches of outer space for brief moments. They were space flights in a narrow technical sense, but they were not sustained space flights. It is one thing to accelerate to a speed of 3,800 km/hr going straight up and peeping above the atmosphere, but it is a far more difficult engineering problem to accelerate a vehicle to the 27,000 km/hr necessary to achieve orbit and, as was tragically demonstrated by the disintegration of the Space Shuttle *Columbia,* to safely decelerate from that speed. Meanwhile, Pan Am is long gone, and there are no plans for a Ho Jo in the sky.

Clearly, technocracy is capable of outperforming technoeconomy in the short run, but its weakness is lack of sustainability. Technocracy can actually throw away capability after it has force-grown it at tremendous taxpayer expense. In the 1960s, the United States government developed the Apollo manned lunar transportation system, the largest launch vehicle system (Saturn V, which suffered not a single failure in 13 launches), and the Nuclear Engine for Rocket Vehicle Applications (NERVA) upper stage for manned interplanetary missions, and within a few years abandoned all of them. These capabilities have yet to be

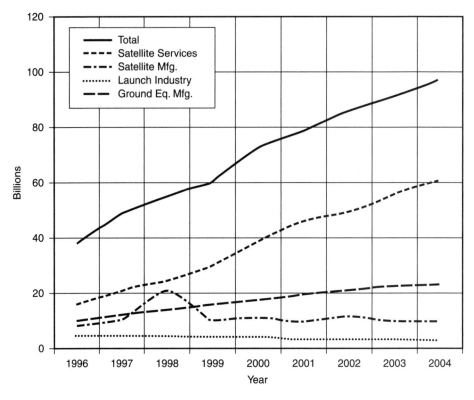

Figure 9.4: U.S. Commercial Space Industry Revenue, 1996–2004
Source: www.sia.org/industry_overview/04industrystats.ppt

duplicated (see Figure 1.1). Since the "Apollo spike," advances in U.S. government-sponsored civil space activities have been incremental. As a theoretical construct, the technocratic spike can be represented as a steep parabola (see Figure 9.4). The period of incremental activity following the spike can be represented as a linear trend. Technoeconomy, on the other hand, builds steadily, compounding its momentum, and can be represented as a geometric progression. One can speculate that in time, private-sector activities and capabilities could catch and surpass public-sector civil activities and capabilities, except that it is difficult to predict when the politically-motivated technocratic spike might occur. Four years after President George W. Bush's announcement of his Vision for Space exploration, as the United States prepares to develop new payload capability to support the return of humans to the Moon and expeditions to Mars, it appears that a new technocratic spike may be in progress (see Figures 9.5 and 9.6). Nope

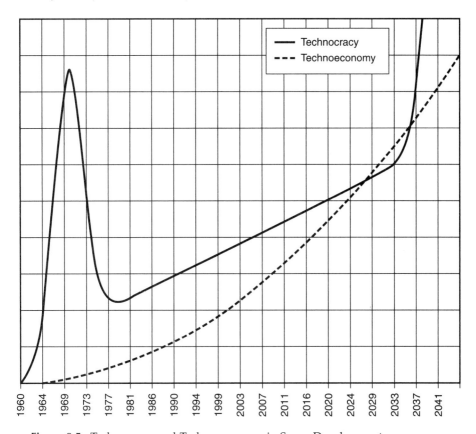

Figure 9.5: Technocracy and Technoeconomy in Space Development

Again, cost is not the primary consideration in political decisions, but it always will be in business decisions. Yet, Declan O'Donnell claims:

> The mass utilization of space resources for commercial and settlement purposes has not yet begun. There are no substantial technical barriers to living and working in space though cost is still a problem. Other problems are delaying our conquest of the final frontier: Space law and space policy problems.[9]

Just because the technical capability to do something exists does not mean that it is a smart business proposition. Moscow, London, New York, San Francisco, and Tokyo all have subway systems, and "there are no substantial technical barriers" to building a subway linking one city to the other, "though cost is still a problem." Cost is hardly a trivial problem, especially in business. It is difficult to imagine what "law and policy problems" might be resolved to make a global subway project a reality.

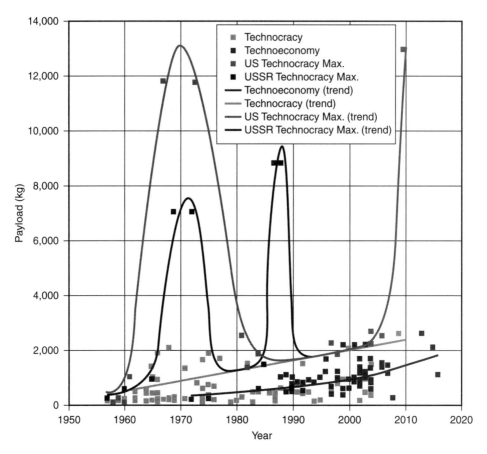

Figure 9.6: Launch Vehicle Payload Capability, Historical Trends

As engineers say, this class of problems is "in the noise level" compared to cost/benefit considerations.

Although there may be regulatory red tape that national governments need to streamline in order to lower barriers to private enterprise in outer space, the hue and cry over extraterrestrial real property rights is a red herring. Opening the floodgates to corporate planetary land grabs would close free access to space that the current international legal regime guarantees. Abrogating the international legal structure that has kept the peace in outer space for four decades would sow the seeds of future interplanetary armed conflict.

The real barrier to commercializing space is the huge capital investment that is required to develop a transplanetary infrastructure. Libertarian space cowboys imagine that private enterprise can pull itself up to the

Moon and Mars by its own bootstraps. These assertions ignore the history of opening frontiers.

In the early days of railroads, a private company might build a line from New York to Buffalo, but New York was already there, Buffalo was already there, and there were Albany and Syracuse in between. On the other hand, the transcontinental railroad that opened the West was a massive U.S. government project to span a vast expanse of nothingness. Similarly, the Russian government built the trans-Siberian railroad.

A French joint-stock company went bankrupt beginning the Panama Canal; the U.S. government stepped in and finished the job. The St. Lawrence Seaway was the joint project of two national governments.

The U.S. government funded the interstate freeway system, which enabled a massive expansion of the automobile industry, trucking industry, the oil industry, and the suburbs.

The airline industry initially developed under federal contracts to transport mail. The Boeing B-707 was developed under an Air Force contract as the KC-135 tanker. The Lockheed L-1011 Tristar development project bankrupted not only Lockheed but Rolls-Royce as well, which was developing the jet engines for that airliner; the U.S. and British governments stepped in to bail out these companies.

The commercial space launchers in service today were all originally developed on government contracts. Most of them began as ballistic missiles to deliver nuclear weapons to distant targets.

Today, private companies build and operate trucks, ships, aircraft, launch vehicles, and satellites, but governments that maintain the highways, seaports, airports, and spaceports—the infrastructure that is the foundation of all of these commercial activities.

Developing infrastructure is a huge capital investment, while maintaining and operating it has a very low profit margin at best. This is something that government is better positioned to do than private enterprise. It has long been recognized that government has a legitimate role "to promote the general welfare" by providing the public goods that enable private goods to flourish.

The libertarian mantra that "government is the problem" is nonsensical. Neither is government the entire solution, but it is a necessary partner in the solution—on land and on sea, in the air and in space. Building a transplanetary infrastructure is not something that private enterprise is going to accomplish, except in the far future. First must come the political vision to build rainbow bridges to the heavens, then will come the economic incentive to travel them.

What makes libertarian rhetoric so seductive is that government seems to have dropped the ball. The golden age of Mercury, Gemini, and Apollo is long gone. During that time, anything seemed possible. It was anticipated that there would be a fully reusable launch system, a space station, a Moon base, and human expeditions to Mars, all by the early 1980s. The technology for all of this was either in hand or within reach, but there was no political necessity, and there certainly was no economic rationale. Clearly, if government were the problem, private enterprise failed to provide a solution. Private enterprise never built a space station or a Moon base, or sent humans to Mars. Is it likely to in the near future?

Government has been getting an increasingly bad rap in the space advocacy community since the end of the Apollo era, but in truth the mad dash to the Moon was unsustainable, and measuring subsequent progress against the Apollo standard reflects unrealistically high expectations. Apollo was a Cold War anomaly that has not been repeated, and that may have no analog in the future.

Again, the central problem is infrastructure. When the Apollo program ended, it left some ground infrastructure (assembly and launch facilities later used by the Space Shuttle program) but no space infrastructure, and in that respect it was a developmental dead end. Political motivation for government to build lasting infrastructure is generated by private sector anticipation of colonizing a new human ecology in which it can produce profit. This is the common thread in all of the aforementioned government infrastructure projects. In contrast, no government has bothered to build a tunnel under the Bering Strait; there are no roads on either side, and so there is little prospect of a sustainable human ecology there. This is not to say that there will never be a Bering Tunnel; it just won't be any time soon.

This may sound like a chicken-and-egg problem. Private enterprise is ill-positioned to develop infrastructure that it requires to thrive. Technocracy—government-directed technological development—has its limits, and may be politically motivated to develop capabilities that have little or no economic utility. A case in point is the depopulation of Siberia that has been occurring since the collapse of Communism. The Soviet Union built infrastructure and forcibly moved population in a massive effort to colonize Siberia and extract its natural resources. Under a command economy, it was not clear that this was an uneconomical project, but as Russia has transitioned to a market economy, an increasing number of people have found that they cannot make a decent living in Siberia, despite its vast natural wealth. There are enormous costs associated with extracting those resources in the extreme environment, and furthermore, there are considerable costs attached to transporting goods out of this remote region of the Earth to market. So, millions of Russians are abandoning the frontier to return to the bosom of Mother Russia's European heartland. Siberia is paradise compared to the distant and forbidding Moon and Mars, yet here private enterprise is retreating from an ecology that government established. Private enterprise only recently duplicated Alan Shepard's 1961 suborbital flight. How credible is it that private enterprise is going to blaze trails to the planets in our lifetime?

It is about as credible as the hype about living on the Moon that baby boomers read in the *Weekly Reader* 40 years ago, or the grand vision of solar power satellite constellations 30 years ago, or a fleet of commercially owned and operated Space Shuttles 20 years ago, or the Iridium mobile telephone satellite constellation 10 years ago. It seems like every time you turn around, space endeavors are being oversold, whether they are governmental or commercial.

However, developing a spacefaring civilization is not an insoluble chicken-and-egg conundrum. It is more subtle than that, and there are solutions—not in all cases, but on the margins. Obviously, progress does occur, and while the pace of progress is not immutable, it does have constraints. The key conceptualization is of government and private enterprise in a push-pull relationship. When private interest becomes curious about what lies over the five-year return-on-investment horizon, it nudges government to stand straight and see further over that horizon. If the vista is promising, private interest encourages government to build the rainbow bridge to the pot of gold. Government then gets its piece of the action by taxing that pot of gold.

The challenge is in recognizing that not every horizon hides a pot of gold, or if it does, it can be too costly to bring it home with the means at hand. Space technology is not a magic wand, and the High Frontier is not the Promised Land. Laissez-faire libertarianism is not the answer to space development any more than command-economy technocracy was. As John Kenneth Galbraith prescribed for the United States half a century ago, what is required is a social balance between public goods and private goods.[10] The concept of and need for sociopolitical balance between various economic power centers in society, including government, corporations, organized labor, international civil society, et cetera, is also described in Raymond Miller's Multicentric Organizational model of political economy.[11] For space development to proceed and to succeed there must be a partnership between government and enterprise as well as among governments and enterprises, a transnational partnership of governmental and nongovernmental entities.[12] It is not merely corporations, but all sectors of human society, that must go into space.

THE TRAGEDY OF THE COSMOS

Some space property rights advocates have latched onto Garrett Hardin's phrase, "The Tragedy of the Commons," to rationalize conferring unrestricted private property rights on celestial bodies.

> Picture a pasture open to all. It is to be expected that each herdsman will try to keep as many cattle as possible on the commons. . . .
>
> As a rational being, each herdsman seeks to maximize his gain. Explicitly or implicitly, more or less consciously, he asks, "What is the utility to me of adding one more animal to my herd?" This utility has one negative and one positive component.
>
> 1. The positive component is a function of the increment of one animal. Since the herdsman receives all the proceeds from the sale of the additional animal, the positive utility is nearly + 1.
> 2. The negative component is a function of the additional overgrazing created by one more animal. Since, however, the effects of overgrazing are shared by all the herdsmen, the negative utility for any particular decision-making herdsman is only a fraction of − 1.
>
> Adding together the component partial utilities, the rational herdsman concludes that the only sensible course for him to pursue is to add another animal to his herd. And another . . . . But this is the conclusion reached by each and every rational herdsman sharing a commons. Therein is the tragedy. Each man is locked into a system that compels him to increase his herd without limit—in a world that is limited. Ruin is the destination toward which all men rush, each pursuing his own best interest in a society that believes in the freedom of the commons. Freedom in a commons brings ruin to all.[13]

Private property rights advocates argue that if each herdsman fenced off a portion of the pasture for his private use, each would manage his property so as to avoid overgrazing, since the negative utility of overgrazing would be borne by the individual landowner. Space libertarians extrapolate this proposition to the Moon and other celestial bodies, asserting that what is owned by none will be plundered by all, once the means to do so become available. They argue that the very existence of a commons inevitably leads to its ruin. But this is taking Hardin out of context; his pastoral scenario is very clearly subtitled, "Tragedy of Freedom in a Commons." It is a thought exercise on the consequences of unrestricted individual freedom, and at that basic level, by carrying it to an extreme, the exercise questions the very thing that libertarians hold most dear.

The "tragedy of the commons" scenario is predicated on the assumption that each individual acts in an individual capacity to maximize his own gain, to the ultimate ruin of all. However, Hardin's thought exercise ought not to be taken beyond its intended purpose. There are several mitigating factors left out of the pastoral commons scenario.

One of these is the concept of diminishing returns. It is assumed that increasing the herd by one results in a positive utility of nearly +1. However, with each herdsman acting in this way, the accessible market for meat might approach saturation; the demand relative to supply might drop, in which case each increment of the herd by one results in a positive utility that is less than that of the preceding increment. It is possible for the positive and negative utilities to approach balance before the commons are overgrazed. As this situation is approached, the probability of the herdsmen coming to a cooperative agreement increases, resulting in a joint management of the commons and a cartel that regulates the sizes of the herds.

This brings up the second factor. Understanding that the individually played scenario results in tragedy triggers a change in human behavior. Humans are not only individuals, they are members in a social group. Humans are not only competitive, they are cooperative. As they see impending tragedy that cannot be solved individually, they are likely to value competition less and cooperation more, and come to a social arrangement in the common interest.

It is this point that Hardin makes in quoting Charles Frankel: "Responsibility is the product of definite social arrangements."[14] Social arrangements, not private ownership. The grazing of private herds on public land is regulated by the government. Fishing is regulated by governments in territorial waters and exclusive economic zones, and by international organizations in the open seas. These are everyday examples of social arrangements that manage the commons responsibly. Private ownership is not in the equation.

Hardin explores a number of devices for tempering human selfishness. One of these is the dreaded "t" word—taxes—which opponents of the Moon Agreement used to defeat it (see chapter five, "Death and Taxes"):

> But temperance also can be created by coercion. Taxing is a good coercive device. To keep downtown shoppers temperate in their use of parking space we introduce parking meters for short periods, and traffic fines for longer ones. We need not actually forbid a citizen to park as long as he wants to; we need merely make it increasingly expensive for him to do so. Not prohibition, but carefully biased options are what we offer him. . . .
>
> To say that we mutually agree to coercion is not to say that we are required to enjoy it, or even to pretend we enjoy it. Who enjoys taxes? We all grumble about them. But we accept compulsory taxes because we recognize that voluntary taxes would favor the conscienceless. We institute and

(grumblingly) support taxes and other coercive devices to escape the horror of the commons.[15]

It is interesting to note that property rights advocates, when appropriating Hardin's "tragedy of the commons" concept for their purposes, invariably ignore his mention of the "t" word. In any case, the main thrust of Hardin's article is that the freedom of human beings to breed, historically an unregulated commons, is no longer appropriate for a planet that is approaching its carrying capacity:

> Perhaps the simplest summary of this analysis of man's population problems is this: the commons, if justifiable at all, is justifiable only under conditions of low-population density. As the human population has increased, the commons has had to be abandoned in one aspect after another.
>
>   The only way we can preserve and nurture other and more precious freedoms is by relinquishing the freedom to breed, and that very soon. "Freedom is the recognition of necessity"—and it is the role of education to reveal to all the necessity of abandoning the freedom to breed. Only so, can we put an end to this aspect of the tragedy of the commons.[16]

The Moon and Mars will have low populations for the foreseeable future, thus it is justifiable to treat them as commons for the foreseeable future. It is only as the population approaches some point at which treatment as commons becomes untenable that new arrangements must be made. Hardin uses the word "commons" in a very restricted sense, in terms of absolute liberty, something that everyone is free to use without restriction. A public resource that is subject to either national or international regulation does not fit this definition. The so-called "tragedy of the commons," therefore, is the consequence, not of the commons *per se*, but of insufficient regulation of the commons when regulation becomes necessary. Unfortunately, those seeking to protect the right to exploit extraterrestrial resources may have contributed to an eventual "tragedy of the commons" in defeating the Moon Agreement:

> The 1967 Space Treaty has embodied only the general principles related to the commitments of the states to avoid a harmful contamination of outer space, including the moon and other celestial bodies (art. IX). The 1979 Agreement makes further steps for protection of the moon and other celestial bodies and, consequently, their resources. Art. VII provides not only for taking measures to prevent the disruption of the existing balance of [the] moon's environment "whether by introducing adverse changes in such environment, its harmful contamination through the introduction of extra-environmental matter or otherwise" (para. 1), but also consideration may be

given to the designation of areas of the moon having special scientific interest as "international scientific preserves for which special protective arrangement are to be agreed" (para. 3). . . .

The common use of the natural resources of celestial bodies implies their rational utilization in conformity with definite criteria, without any discrimination whatsoever, in the basis of equality of the states and prohibition of any exceptional rights in favour of any of the states. Rational utilization of these resources must be carried on by taking account of the interests of both the international community as a whole and individual states, in the interests of not only the present but the future generations as well.[17]

Since outer space is a commons by international law, and the overturning of that status is exceedingly unlikely, the only practical option for forestalling the "tragedy of the commons" is to establish a governing regime for managing the rational utilization of resources, for it is unrestricted liberty that inevitably leads to tragedy. However, in most cases into the foreseeable future, it would be a mistake for such a regime to be overly constraining.

The Tragedy of the Commons does not readily apply to space resources. Strict administration of scarce resources makes sense in the case of geostationary orbits but the vast resources of space and celestial bodies simply stagger the mind.[18]

INTERPLANETARY TERMS OF TRADE

The case for the celestial bodies being raw materials sources for Earth-based industry is weak due to the high cost of interplanetary transportation for the foreseeable future. Trade with Earth will have to await a drastic reduction in transportation costs. The most credible use of celestial resources is for either local use or use elsewhere in space. In the absence of commercial revenue from direct trade with Earth, interplanetary trade not involving Earth directly would necessarily involve government contracts to private companies to provide goods and services to government-owned operations. Earth is the source of investment capital for outer space, and the accumulation of hard currency in outer space will require the creation and exchange of something that is of value to Earth. It is such productive activities that must form the basis of a largely commercial, nongovernmental, extraterrestrial economy.

A relatively early prospect for overcoming the transportation barrier is helium-3 on the Moon, which is often cited as a raw material that could be economically transported to Earth; however, the distribution,

Table 9.2: Absolute Advantage

	Production Rates		
	Planet A **(units/worker)**	**Planet B** **(units/worker)**	
Commodity 1	250	150	
Commodity 2	100	200	
	Production without Trade		
	Planet A **(units)**	**Planet B** **(units)**	**Total** **(units)**
Commodity 1	25,000,000	15,000,000	40,000,000
Commodity 2	10,000,000	20,000,000	30,000,000
	Production with Trade		
	Planet A **(units)**	**Planet B** **(units)**	**Total** **(units)**
Commodity 1	50,000,000	0	50,000,000
Commodity 2	0	40,000,000	40,000,000

abundance, and availability of helium-3 is not well characterized, and the fusion power technology on which the commercial value of helium-3 is predicated may not be available for 20 to 50 years. Another possibility is that lunar material might be the most economical source of materials for construction of solar power satellites that would transmit energy to Earth on a commercial basis.

As commercial interplanetary trade develops, interesting avenues of speculation arise. Adam Smith explained how two countries could profit in trading with each other by using their absolute advantages over each other in producing different commodities.[19] For instance, consider two planets, A and B, each capable of producing commodities 1 and 2 (see Table 9.2). With production time held constant and neglecting transportation costs, on Planet A, the average worker is capable of producing 250 units of Commodity 1, whereas on Planet B, the average worker is capable of producing 150 units of Commodity 1. Planet A enjoys an absolute advantage over Planet B in Commodity 1, since its workers are more efficient in producing that commodity. On the other hand, on Planet A, the average worker is capable of producing 100 units of Commodity 2, whereas on Planet B, the average worker is capable of producing 200 units of Commodity 2. Planet B enjoys an absolute advantage over Planet A in Commodity 2. If Planet A and Planet B do not trade, and each divides its workforce of 200,000 people in half to produce each commodity, they will produce 40 million units of Commodity 1 and

Table 9.3: Comparative Advantage

	Planet A (units)	Planet B (units)
Commodity 3	1,200,000	200,000
Commodity 4	300,000	100,000

30 million units of Commodity 2. However, if Planet A and Planet B do trade, they can specialize in producing the commodity in which they enjoy absolute advantage, thereby producing 50 million units of Commodity 1 and 40 million units of Commodity 2. Both planets profit from trade. Each is buying a cheaper commodity from the other than it could produce. Productivity increases on both planets.

Now suppose that on Planet A the average worker is capable of producing 1,200 units of Commodity 3, whereas on Planet B, the average worker is capable of producing 200 units of Commodity 3 (see Table 9.3). Planet A enjoys an absolute advantage over Planet B in Commodity 3. Also, on Planet A, the average worker is capable of producing 300 units of Commodity 4, whereas on Planet B, the average worker is capable of producing 100 units of Commodity 4. Planet A enjoys an absolute advantage over Planet B in Commodity 4.

Planet A is more efficient at producing both commodities. Why should it trade with Planet B? Planet A would lose the opportunity to produce four units of Commodity 3 to produce one unit of Commodity 4, but Planet B would lose the opportunity to produce only two units of Commodity 3 to produce one unit of Commodity 4. It is relatively cheaper for Planet A to import Planet B's Commodity 4, because it minimizes the lost opportunity to produce Commodity 3. It is the production ratios within the two economies that make comparative advantage work. Planet B also profits from trading with Planet A, since Planet A's Commodity 3 is cheaper than Planet B's Commodity 3.

David Ricardo's model of comparative advantage was valid on Earth in the days when capital was for the most part restricted to the borders of one country.[20] It does not work the same way when capital is mobile as it is today. Capital moves to where labor is cheapest, thus jobs are lost, and the displaced workers may not find work in other industries. In any case, absolute and comparative advantages can fluctuate with the cost of materials, cost of capital, cost of labor, skill level, and other factors.

Meanwhile, there is an additional complexity when it comes to interplanetary trade. Extremely problematic environmental factors such as high atmospheric pressure (Venus and the gas giants) or extremely long travel times (the gas giants' moons) will probably delay the establishment of settlement on these planets indefinitely. The remaining list of celestial

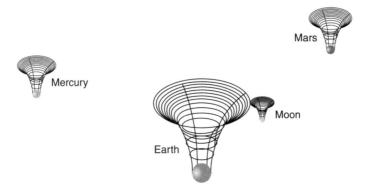

Figure 9.7: Gravity Wells

bodies on which settlements might be established are Mercury, the Moon, Mars and its satellites, and various asteroids. These bodies have one thing in common: they are smaller, less massive than Earth. Because of this, they have less powerful gravitational fields, thus it takes less fuel to land on and launch from them, which makes them less expensive ports for moving goods. This translates as a permanent, gravitational advantage. Nature imposes a permanent, nonreciprocal tariff on terrestrial goods shipped to these bodies. As a result, it will be less expensive for extraterrestrial settlements to ship goods to each other and to Earth than it will be for Earth to ship goods to them. There may be persistent trade deficits/surpluses between planets. This situation has some analogues on Earth; for instance, a landlocked country without navigable rivers would need to go to the expense of building railroads and highways to access a seaport in a neighboring country. But these relative differences in transshipment costs are negligible in comparison with the differences between the planets in their gravity wells (see Figure 9.7).

To begin the analysis, we use Konstantin E. Tsiolkovskij's rocket equation:

$$v = v_e \ln \frac{m_0}{m_1}$$

where:

  m_0 = initial total mass

  m_1 = final total mass

  Δv = integration over time of the magnitude of the acceleration produced by using the rocket engine

  v_e = velocity of the rocket exhaust with respect to the rocket (the specific impulse, or, if measured in time, that multiplied by gravity-on-Earth acceleration)

Table 9.4: Surface Gravity and Escape Velocity

	Surface Gravity (m/s²)	Escape Velocity (km/s)
Mercury	3.701	4.435
Earth	9.7801	11.186
Moon	1.622	2.38
Mars	3.69	5.027

Solving for m_1, the final total mass:

$$m_1 = m_0 e \, \Delta v \, / \, \Delta v_e$$

Let us assume that the same rocket propellants are used everywhere, thus v_e can be held constant. The ratio of the masses that can be propelled from one planet to another with the same amount of propellant is thus dependent on the difference in their escape velocities (see Table 9.4) and is given by:

$$\frac{m_{p1}}{m_{p2}} = e(\Delta v_{p2} - \Delta v_{p1})$$

This equation gives a first approximation of the gravitational advantage of one planet over another. Planet-to-planet values are given in Table 9.5. Second order terms would include taking into consideration such things as the relative expense of descent modes (aerobraking, parachute, and braking propulsion for Earth and Mars, braking propulsion only for the Moon and Mercury). Another consideration to bear in mind is that the ratios in Table 9.5 assume the same propulsion systems throughout the system.

The Moon's gravitational advantage might be enhanced by the use of mass drivers, a form of electromagnetic propulsion rather than chemical propulsion. Also, the development of scramjet technology could mitigate Earth's gravitational disadvantage.

Table 9.5: Gravitational Advantage

To → From ↓	Mercury	Earth	Moon	Mars
Mercury		854.91	0.12809	1.8076
Earth	0.00117		0.00015	0.0021
Moon	7.80684	6674.17		14.1116
Mars	0.55322	472.96	0.07086	

The economic rationale often cited for settling on other celestial bodies is that they are future sources of primary resources. Let us look forward to a distant time when low transportation costs make this economically feasible. In the past, colonization led to exploitative relationships, in which colonial powers took advantage of cheap, unskilled labor in the colonies to extract raw materials, and sell high value added-goods manufactured by skilled labor back to the colonies. Although the European colonial empires were dismantled in the mid-twentieth century, residual trade relationships have persisted into the twenty-first century. Immanuel Wallerstein refers to the industrial core of the capitalist world system, the semi-periphery of lesser industrialized states, and the underdeveloped periphery.[21] However, space settlements will be high-tech by their very nature, and will be populated by highly educated, highly skilled workforces. Thus, once settlements are able to provide for their own subsistence, they will be able to turn to high value-added productive activities for export, many of which will be competitive with terrestrial products elsewhere in the solar system due to gravitational advantage.

THE FUTURE OF THE NATION-STATE SYSTEM

Medieval Europe was characterized by the feudal mode of production. Sovereignty was dynastic, property belonged to hereditary nobilities, and productive labor was performed by serfs who were bound to the land. The overarching power was the Catholic Church, including its secular subsidiary, the Holy Roman Empire. The rise of the capitalist mode of production in towns and the interstate trade between them created new sources and stores of wealth rivaling those of the landed nobility. The rise of city-states and city-leagues challenged the power of the dynastic states and the Empire. The rise of literacy in the growing cities challenged the religious authority of the Church. An important result of the 1648 Peace of Westphalia that ended the Thirty Years' War was that it laid to rest the idea of the Holy Roman Empire having secular dominion over much of Western Christendom. The loss of this overarching structure created the necessity for feudal entities to coalesce into larger, more centralized states to defend against each other. Imperial free cities were no longer adequately protected by the Empire, and joined the larger states that were forming. Hendrik Spruyt, Harris Professor of International Relations at Northwestern University, describes the ascendance of the nation-state over medieval forms of sovereignty:

> At the end of the feudal era, a dramatic economic change occurred. Localized barter exchange started to give way to monetary exchange and translocal

trade. By the beginning of the fourteenth century, a variety of new institutional forms had emerged for organizing political and economic life. Sovereign territorial states, city-leagues, and city-states all tried to tap into the new sources of economic wealth, particularly long-distance trade. Indeed, the city-based political organizations initially did very well. In the long run, however, roughly by the middle of the seventeenth century, city-states and city-leagues had fallen by the wayside. . . .

I argue that the sovereign territorial state prevailed because it proved more effective at preventing defection by its members, reducing internal transaction costs, and making credible commitments to other units. . . . [S]overeign rulers were better at centralizing jurisdiction and authority. Consequently, they were in a better position to prevent free riding and to gradually rationalize their economies and standardize coinage and weights and measures. This economic rationalization corresponded with a greater capacity to wage war. The institutional makeup of sovereign territorial states thus gave them competitive advantages over other organizational possibilities.[22]

The so-called nation-state system was born. However, Spruyt's use of the more general term "sovereign territorial state" is more appropriate.

A nation-state is a specific form of state (a political entity), which exists to provide a sovereign territory for a particular nation (a cultural entity), and which derives its legitimacy from that function. . . . Typically it is a unitary state with a single system of law and government. It is almost by definition a sovereign state, meaning that there is no external authority above the state itself.

. . . The nation-state implies the parallel occurrence of a state and a nation. In the ideal nation-state, the population consists of the nation and only of the nation: the state not only houses it, but protects it and its national identity (i.e., they coincide exactly): every member of the nation is a permanent resident of the nation-state, and no member of the nation permanently resides outside it. There are no ideal nation-states, but examples of near ideal nation-states might include Japan and Iceland. This ideal has influenced almost all existing sovereign states, and they cannot be understood without reference to that model. . . . Thus, the term nation-state is also used, imprecisely, for a state that attempts to promote a single national identity, often beginning with a single national language.[23]

There are many states in today's international system that are not nation-states in the strict sense. At one end of the spectrum are the principalities of Liechtenstein, Andorra, and Monaco, and the republics of San Marino and Singapore; these are essentially city-states. Numerous territorial states contain many national groups, the United Kingdom and the Russian Federation being two examples. Switzerland, whose independence was recognized in the Peace of Westphalia, has no common language. The United States has historically boasted of being the "melting pot" of cultures. However, in keeping with general usage, sovereign territorial states will continue to be referred to as "nation-states," and the system of relations between them as the "international system" or the "nation-state system."

Marxist ideology predicts the ultimate demise of the Westphalian system, to be replaced by an international workers' union; however, the capitalist world system has outlived the centrally planned economic system that purported to be founded on Marxist principles. Ironically, capitalist globalization has often been cited as a process that threatens the sovereignty of the state, and therefore has great implications for the future character of the nation-state system. By the late twentieth century, transnational production systems had enhanced the ability of corporations to bypass the nationally based political influence of trade unions and environmental movements. In the view of many scholars, the various interrelated processes of globalization, which are making national borders more porous to trade in goods and services, to capital, and to information, including sociopolitical ideas and norms, are rendering the nation-state increasingly less relevant. Robert Gilpin, professor of public and international affairs at Princeton University, observes:

> As a consequence, multinational firms have become extremely important in determining the economic, political, and social welfare of many nations. Controlling much of the world's investment capital, technology, and access to global markets, such firms have become major players not only in international economic, but political affairs as well.[24]

Manfred Steger, professor of global studies at the Royal Melbourne Institute of Technology, writes:

> Most of the debate on political globalization involves the weighing of conflicting evidence with regard to the fate of the modern nation-state. In particular, two questions have moved to the top of the research agenda. First, what are the political causes for the massive flows of capital, money, and technology across territorial boundaries? Second, do these flows constitute a serious challenge to the power of the nation-state? These questions imply that economic globalization might be leading to the reduced control of national governments over economic policy. The latter question, in particular, involves an important subset of issues pertaining to the principle of state sovereignty, the growing impact of intergovernmental organizations, and the prospects for global governance.[25]

One influential group of scholars sees politics as being "rendered almost powerless in the face of an unstoppable and irreversible technoeconomic juggernaut that will crush all governmental attempts to reintroduce restrictive policies and regulations." And, not only are political institutions no longer in control of national economies, but now economic interests are in control of national policies. According to this view, we are now "in a new phase in world history in which the role of government will be reduced to that of a handmaiden to free-market forces."[26] Not that this is a particularly novel insight, for Karl Marx characterized government as the handmaiden of the bourgeoisie a century and a half ago, as the nineteenth century era of globalization was in full swing. In any case, in the most extreme expression of this view, national borders will ultimately cease to be meaningful concepts, and the fate of the planet will be entirely in the hands of global capitalist forces.

Meanwhile, the propensity of the nation-state system to fence off peoples from each other and to organize them for aggressive purposes has been remarked on by many. The unparalleled destructiveness of the twentieth century wars led some to question the nation-state's ability to deliver on a state's most basic *raison d'être*: security. European leaders in particular have been crafting a system of political integration that is transcending the nation-state. North Atlantic Treaty Organization (NATO) secretary-general Javier Solana states that "humanity and democracy [were] two principles essentially irrelevant to the original Westphalian order. . . . [T]he principle of sovereignty it relied on . . . produced the basis for rivalry, not community of states; exclusion, not integration."[27] German Foreign Minister Joschka Fischer has declared the Westphalian system to be obsolete:

> The core of the concept of Europe after 1945 was and still is a rejection of the European balance-of-power principle and the hegemonic ambitions of individual states that had emerged following the Peace of Westphalia in 1648, a rejection which took the form of closer meshing of vital interests and the transfer of nation-state sovereign rights to supranational European institutions.[28]

The "Clash of Civilizations" envisioned by Samuel P. Huntington as manifested in the War on Terrorism has given rise to a new threat to the international system. Following the March 11, 2004, terror bombings in Madrid, a self-proclaimed al-Qaeda spokesperson declared that "the international system built-up by the West since the Treaty of Westphalia will collapse; and a new international system will rise under the leadership of a mighty Islamic state."[29] On the other hand, some argue that the Westphalian system collapsed with the disintegration of the Soviet Union, which left the United States as the global hegemon. Certainly Yugoslavia in the late 1990s had, and Afghanistan and Iraq in the first years of the twenty-first century have, whatever sovereignty the sole superpower has suffered them to exercise.

Jorg Friedrichs, Weber Fellow at the European University Institute, offers a novel perspective on the complex and fluid state of affairs. He notes that neither conventional international relations theory nor the discourse about globalization seem able to account for the apparent contradictions between globalization, fragmentation, and sovereign statehood. As a conceptual alternative, he introduces the idea of "new medievalism," in the sense that today's system is characterized by overlapping authority and multiple loyalty, held together by a duality of competing universalistic claims, much as:

> the Middle Ages were characterized by a highly fragmented and decentralized network of sociopolitical relationships, held together by the competing universalistic claims of the Empire and the Church. In an analogous way, the post-international world is characterized by a complicated web of societal identities, held together by the antagonistic organizational claims of the nation-state system and the transnational market economy. New medievalism provides a conceptual synthesis which hopefully transcends some of the current deadlocks of IR theory and, at the same time, goes beyond the fundamental limitations of the globalization discourse.[30]

Lifting our gaze skyward, the question arises, what forms of governance shall we take with us, and what shall we leave behind? We have an opportunity for a revolution in human thought regarding human nature and human rights, the relationship of individual, society, and state, and to question the factors of production, the means of production, and modes of production. In short, as one writer on outer space law remarked to me, we get to do the Enlightenment over again.

> We can challenge sociology to provide the conceptual basis of the social design for space settlement, to construct social models required for the successful adaptation of Humankind to space. Planning the social design of space settlements is necessary for the success of space industrialization; without appropriate social organization, accomplishment of industrialization tasks will fail.
>
> Rational planning for the social design of space settlements must rest on the knowledge of social scientists and scholars. This requirement presents a particularly difficult problem for those who plan the social aspects of space settlements. This does not constitute an insurmountable difficulty, however, because knowledge of human social responses to various types of problems and difficulties is available; this knowledge can be applied to analysis of the unique problems that are likely to require societal resolution in space settlements from which logical models of social design can be developed. These

logical models can serve as a starting point for social planning and can be modified as additional knowledge and insight are developed.[31]

What were the forces and conditions present in the seventeenth and eighteenth centuries that set the Enlightenment in motion? Are there analogous forces and conditions present in the twenty-first century that might set in motion a new Enlightenment?

From the early sixteenth to the mid-seventeenth centuries, Europe was ravaged by religious wars and civil wars as the Protestant Reformation broke the temporal power of the Catholic Church, as well as its hold over knowledge and learning. Stepping into the voids left by the fading of the feudal order and the Holy Roman Empire were the nation-state political system, the capitalist production system, and the scientific knowledge system. The new order solved old problems and created new ones. There were still wars, persecutions, and injustices, and if anything, the problem of war was exacerbated. Whereas feudal lords were rarely able to raise armies of more than a few tens of thousands, a particular advantage of the nation-state was its ability to mobilize millions in the machinery of death, and its principal evil was its propensity to set that machine in motion again and again.

The Enlightenment occurred just as a new ecological opportunity was opening for the Westphalian system. European colonies in the New World were reaching a level of development at which they could challenge their mother countries and gain independence. These new American nation-states were distinctly different in character from their European antecedents; whereas most of Europe was ruled by hereditary monarchs, the American states were theoretically republics.* With varying degrees of success, the ideals of the Enlightenment were expressed in the political entities that sprouted from the new soil of the Americas. A century and half after its birth, the strongest of these American republics was able to intervene decisively in a war that was devastating Europe, to do so again a generation later to defeat one form of totalitarianism, and to protect half of the continent for two more generations from another form of totalitarianism. Today, in its continuing lead role in NATO, the United States acts as the guarantor of European stability as the continent's institutions expand their membership and deepen their level of integration. Europe is a union of peaceful, liberal, democratic states.

Here on Earth, we face in the twenty-first century a possible receding of the nation-state system as international borders become more porous to

---

* The Empire of Brazil, 1822–1889, proclaimed by Pedro I and later ruled by his son Pedro II, was a notable exception.

trade, travel, and information. On top of this, humankind stands on the threshold of expanding into the new ecologies of the solar system. The nation-state does not exist out there, and the current body of international law makes it doubtful that it ever will. Since Article 2 of the Outer Space Treaty Outer prohibits "national appropriation by claim of sovereignty," it seems that there can be no national sovereignty. As strange as it may sound, it appears that the nation-state has chosen not to extend itself into the cosmos, but to restrict itself to Earth.

> What is the reason for [the] renunciation of the right to the occupation of a
> *res nullius* for the particular benefit of one state? It might be judicial—a more
> lofty concept of justice and of international solidarity. It might be
> economic—the enormous cost of astronautic enterprises, which would
> make it difficult for a single nation to sustain the expenses involved.
>
> It is almost certain that the reason is political; the two principal protag-
> onists in international politics today, the United States and Russia, do not
> yet know with certainty who will be the first to arrive and, mutually fear-
> ing the result of a triumph by the adversary, are trying to bring all the
> nations into the game, thus making a common cause with the winner and
> reducing the risks.
>
> It is vital, however, not to discard the possibility that this is due to an
> awakening of the universal legal conscience to the possibility of a cultural,
> economic and, in certain cases, political unification.[32]

There may very well be states in space of some sort, in that there will need to be some form of self-governance. In some way, shape, or form, there must be human mechanisms to enact law, to interpret law, and to execute law. Someone has to make decisions and to be accountable for them. The buck has to stop with someone.

> The space treaties now in force have not abolished sovereignty or sovereign
> rights in outer space. The relevant Article II of the Outer Space Treaty only
> refers to sovereignty in connection with a ban on national appropriation.
> Nothing is said about prohibiting other forms or expressions of sovereignty
> or sovereign rights. In fact, the Outer Space Treaty provides that states shall
> bear international responsibility for national activities in outer space irre-
> spective of whether such activities are carried out by governmental or non-
> governmental entities. This provision presupposes exercise of some form of
> sovereign rights, jurisdiction or control by the state whose nationals are part
> of the space colony. Otherwise, it is hard to see how a state could be held
> internationally liable for activities over which it had absolutely no control.[33]

However, what degree of sovereignty the space states may have in relation to the nation-states system on Earth, or in relation to each other, is an open question. There will be no nations in space. Nations are defined by distinct ethnicities, languages, and cultures, originating, developing, and largely contained within distinct geographic areas. How distinct will be one lunar settlement from another, or from a Martian settlement? Sovereignty will be localized, and the shape that sovereignty takes will be influenced by the social organization of these settlements. The conceptual groundwork for such organization can and should begin before significant numbers of people are working in space on a regular basis. A quarter-century ago, Niagara University sociologist Stewart B. Whitney called for social scientists to seize the opportunity to envision a new form of society:

> The relatively new idea of space settlements has attracted the interest of few social scientists; publications are rare and appear in disparate sources. Much of the material that is published on "space colonies," "space humanization," "human communities in space, etc., is written by technologists and reveals little significant knowledge about the complexities of human society. . . .
>
> We need an international cadre of space planners. Rather than replicating and concretizing political economic forms of the past through haphazard and string-budget planning efforts . . . space warrants international planning and organization. I am calling for . . . a redevelopment and [re]arrangement of Humankind's priorities: a 21st century political economy.[34]

Jurisdiction will extend only over territory being used for productive activity. Forms of local sovereignty will develop under the international treaty regime. Extraterrestrial settlements will start out as international projects in many cases, and will gradually develop self-governance, but not complete sovereignty. Settlements may develop within a confederation or an integrated federal system, or alternatively, sovereignty may be vested in entities resembling city-states, and possibly city-leagues of voluntarily shared sovereignty will develop.

Since any resources not being used are the commons, and according to Article 1 of the Outer Space Treaty, "there shall be free access to all areas of celestial bodies," there can be no fencing off of large areas for exclusion. Since no large areas can be appropriated, no large areas need be defended against encroachment. Thus there will be no need of a military force, and no need of a strong central government to support and organize such a force. Sovereignty will be vested in the people, and aggregated only to a local level as necessary for self-governance to provide police, fire, health, and infrastructure services for each settlement. Of course, wherever there is interstate commerce there is the potential for piracy, and an interplanetary police force may be necessary to suppress that evil.

In contrast, the security requirements to defend the vast tracts of claimed territory proposed by Alan Wasser would be considerable.

> If we could get something like [the Space Settlement Prize Act] enacted into U.S., and preferably international, law the space race would quickly resume, this time among consortia of private companies. After the first announcement of an attempt to set up a lunar base, others, all over the world, would say, "we can't let them claim the Moon, WE must get there first." Fear of competitors is still the best motivator.[35]

Fear is also the best motivator for violence, especially when alloyed with greed. Assuming that lunar property were to become suddenly lucrative as Wasser envisions. Since his scheme depends on abrogating or ignoring the Outer Space Treaty, there would be no international legal regime to govern this lunar land rush. Governments would recognize the claims of their nationals over the claims of foreigners. One can envision that in the absence of a coherent legal regime, corporate security squads would battle each other to assert conflicting claims, and if the stakes were high enough, governments would be drawn into the fray. National military power always tends to follow the private interests of its citizens and corporations. If Wasser's vision were successful, there would be mercantile wars on the Moon and Mars. Wayne N. White observes:

> The entire history of Earth is one long tale of military conflict over disputed territory, or even outright seizure of territory by governments with no lawful claim to the territory in question. Permitting national claims of territorial sovereignty in Outer Space would only perpetuate that history of conflict.[36]

Historically, private appropriation on a scale of national appropriation has been as fruitful a cause of war as national appropriation itself. As the nation-state system emerged beginning in the mid-seventeenth century, a distinction came to be drawn between sovereign territory and private land. In feudal times, however, there was little or no distinction; the territories of the nation were the king's private lands, or of his vassals. The object of war was to increase the land holdings of the king and the nobility. In the absence of nation-state government, warfare on a corporate-feudal Moon and Mars would be endemic. But of course, because of these concerns, Wasser's vision cannot be successful. Private enterprise will not invest in space in the absence of a stable international legal regime. Going into space is expensive and dangerous enough as it is without having to worry about claim-jumpers, bushwhackers, and privateers.

It might be argued that populations in isolation on these far-flung worlds will develop distinct cultures over time, that new national identities will form, just as they did in the New World, and thus the evils of the Westphalian system will proliferate throughout the solar system. Militating against this, however, is the fact that these entities will not have developed as isolated appendages of mercantile empires; the capitalist multiworld-system will bind them together into a web of production chains and trading relationships. Also, whereas communication between American colonies and Europe took months, and even between colonies took weeks, information will flash across the inner solar system in a matter of minutes. Marshal McLuhan's "global village" will become the "solar system village."[37]

The basic unit of the interplanetary political system may evolve over time, from individual settlements to geographically larger agglomerations, but there will probably remain an overarching regime, such as is envisioned in the Moon Agreement (see chapter five, "Establishment of a Governing Regime"). This regime may gradually evolve into an interplanetary "government," but with much restricted powers compared to those associated with nation-states. Friedrichs's new medievalism might well characterize the governance of the solar system, a complicated web of societal identities, held together by the competing organizational claims of the interplanetary market economy, the interplanetary regime established by treaty, and localized sovereignty. Yet the term "medievalism" has the connotation of something very backward, and in a way, there will be a backward motion as well as a forward motion, but it will be a backward motion informed by forward thinking. Whereas the Reformation led to the agglomeration of sovereignty from dynastic states to the national level, the devolution of sovereignty to the local level and to the individual would better express the ideals of the Enlightenment. Marco G. Markoff exhorts:

> Space lawyers have to help developing the new legal order in space and on celestial bodies. They should endeavour to prevent space contamination with national rivalries and disputes. Instead of acting like dinosaurs in [the] space age, jurists have to impose upon politics an evolution towards a real common use and exploitation of the resources of space and celestial bodies, in the interests and for the benefit of all countries. A new pattern for a better future on the Earth, too, will be created therewith, in full harmony with the positive international law of space.[38]

A contradiction in the outer space private property rights movement is that it is ideologically based in economic liberalism; however, the methods it advocates are realist, in that it dismisses the importance of treaties, advocating their contravention or U.S. withdrawal from them. Global liberalism requires a stable international legal regime, and space liberalism will also.

To be successful, it needs to be institutional in its approach. Another contradiction in the space libertarian schemes for private property rights is that in espousing the assertion of the nation-state over the international community, the effect is to extend the nation-state into outer space, to the detriment of local sovereignty. In contrast, the common law regime of the Regency of United Societies in Space (see chapter three, "Fanfare for the Common Law") would obviate much of the need for sovereign regimes.

In outer space, where nation-states do not and cannot exist, a new paradigm of sovereignty will have an unfettered opportunity to take root and to flourish. We may see representative government as known on Earth give ground to something approaching the Athenian model of direct democracy. Settlements will start out small enough for each adult to participate in the political decision-making process without the need for an elected legislature of representatives to mediate. As these settlements grow, the continuing development of information technology may enable individuals in these highly educated populations to process more of the types of information pertinent to political decisions. A few centuries from now, whatever new forms of governance, new conceptions of rights, and new norms of state behavior that have been invented on the Moon, Mars, and elsewhere in the solar system, may be introduced to "old Earth."

THE FUTURE ECOLOGY OF CAPITALISM

A specter is haunting the solar system.

Whereas progressives blame the voracious appetite of the capitalist system for accelerating the depletion of the world's resources and the destruction of the environment, economic liberals and enviro-capitalists claim that "wealthier is healthier," that more citizens of more affluent societies demand a better environment in which to live. In 1991, economists first reported a systematic relationship between income changes and environmental quality, known as the Environmental Kuznets Curve (EKC). When first unveiled, EKCs revealed a surprising outcome: Some important indicators of environmental quality such as the levels of sulfur dioxide and particulates in the air actually improved as incomes and levels of consumption went up.[39] On the other hand, a recent study concluded that there is little empirical support for an inverted U-shaped relationship between several important air pollutants and national income.[40]

New technologies may alleviate some of the problems. The capitalist drive for efficiency may save it from consuming the entire planet and drive it toward developing more efficient ways of using the limited resources of the planet. Economic liberals argue that successful corporations can afford to invest in more environmentally friendly technologies, but it is not clear that they have the incentive to do so. This may require melding environmental policy and market forces in such a way that the environmental costs of manufacturing a product are factored into its sale price, and the environmental cost of operating a product is amortized over its life cycle. It may also require a more environmentally conscious consumerism that alters the forces of the marketplace to favor products with smaller environmental footprints. Even in the best case scenario, there will probably be significant costs to changing the essential structure of the economic system. It is not clear how we get there from here, who makes the decisions, and how the costs are distributed.

There are limits to growth; however, these are not hard and fast, but are defined in terms of market and social forces and technology. There is a limit to the number of internal combustion automobiles that can be operated on this planet for a given environmental load. Fuel cell technology, being much cleaner, will increase the limit on automobiles; however, the electrical power needed to electrolyze water into hydrogen and oxygen will probably still be supplied by fossil fuel–burning plants for the next several decades. Fusion power generation will increase our limits even further; however, 25 years ago, it was estimated that the commercial application of this technology was 20 to 50 years away, and unfortunately, that is still the estimate.

It is true that we only have one Earth, yet we need not be limited to it forever. In the past, the political left has regarded the space program as "macho and polluting."[41] This is unfortunate, for many advocates for a vigorous space program are socially progressive and environmentally conscious. More to the point, for decades now, remote sensing from orbit has allowed us to discover new resources and manage them more intelligently, and to monitor environmental degradation. Environmentalist Stewart Brand labeled the space satellite "an engine of the ecology movement."[42]

Looking forward, the development of extraterrestrial resources will certainly not provide complete solutions to overpopulation, overconsumption, resource depletion, and environmental degradation, but it will increasingly become part of the solution. The resources of the solar system are not available to us with current chemical propulsion technology, nor does this technology make it feasible for any but a handful of people to emigrate from Earth. Nuclear thermal propulsion (NTP), although more efficient, would engender much of the same social resistance as nuclear fission power plants. (NTP technology was developed in the 1960s for human interplanetary missions, but has never been used in actual spaceflight.) Even so, NTP would probably enable only limited extraterrestrial resource utilization, perhaps employing a few thousand or tens of thousands of off-planet workers. The dramatic breakout will probably have to wait for fusion propulsion technology, which will probably lag the development of Earth-based commercial fusion power generation by several decades. Fusion propulsion, of course, is only a gleam in the eyes of engineers, and its implications cannot yet be well understood. Potentially, however, it could enable the full flowering of a transplanetary economy in which tens of millions of people live and work off-planet in its initial stages. While it is doubtful that even this level of technology will enable enough people to emigrate from Earth to reduce its population by any appreciable degree, it should make it economically feasible to move a considerable portion of environmentally destructive industries off-planet. The Moon, Mars, and the asteroids will become the new economic periphery, as corelike, predominately low-pollution, economic activity expands to encompass most of the Earth (the ultimate EKC). No doubt, new socioeconomic and political problems will arise in this multiplanetary venue, even as we solve some of our old ones. In any case, it is possible that the crisis in capitalism that Wallerstein foresees might either be averted or informed by economic expansion into the solar system, if we can keep the world economic system from hitting the wall before then.

Whereas mainstream economics takes the market economy as a given, Wallerstein points out that capitalism is a historical phenomenon; it had a beginning, and presumably it will have an end. It played a role in transforming medievalism into the Westphalian system and spawned the Industrial Revolution. Nation-state policy interaction with capitalism has yielded in turn mercantilism, trade liberalism, imperialism, fascism, and neoliberalism. And, capitalism has morphed with each of these political changes, adapting to the political environment. All of these changes have occurred within the lifespan of the Westphalian system. It may not be possible to predict what capitalism will morph into under an interplanetary political environment characterized by new medievalism, but it is certainly possible to predict that it will morph. Given the many phases it has experienced within the Westphalian system, the outcome of its transformation under new medievalism might well be unrecognizable to us. This suggests that we are in little danger of approaching the "End of History" any time soon, that there is more socioeconomic evolution to come as the human species colonizes new ecological niches in the cosmos. There will be new struggles over modes of production and political ideology, but we may hope that the history of the future will be far less bloody.

NOTES

1. Futron Corporation. 2002 (September 6). "Space Transportation Costs: Trends in Price Per Pound to Orbit, 1990–2000." Available from http://www. futron.com/pdf/FutronLaunchCostWP.pdf; accessed January 21, 2006.

2. White, Wayne N., e-mail message to author, January 22, 2006.

3. White, Wayne N., e-mail message to author, January 23, 2006.

4. National Aeronautics and Space Administration, Glenn Research Center. 2006. "Specific Impulse." Available from http://www.grc.nasa.gov/WWW/ K-12/airplane/specimp.html; accessed February 23, 2006.

5. Kucinich, Dennis. 2004. "Kucinich Says NASA's Budget Should Be Tripled." *Space Daily*, 9 May. Available from http://www.spacedaily.com/news/ nasa-04j.html; accessed January 30, 2006.

6. Cook, Rhodes. 2004. "2004 Nationwide Democratic Primary Results and Delegate Count." Available from http://www.rhodescook.com/primary. analysis.html; accessed January 6, 2006.

7. Anderson, Poul. 1963 (June). "Territory." *Analog*.

8. Nolan, Patrick, and Gerhard Lenski. 2006. *Human Societies*. Boulder: Paradigm, 350.

9. O'Donnell, Declan J. 1999. "Property Rights and Space Resources Development." Available from http://www.mines.edu/research/srr/ODonnell.pdf; accessed February 1, 2005.

10. Sackrey, Charles, and Geoffrey Schneider with Janet Knoedler. 2002. "John Kenneth Galbraith and the Theory of Social Balance." In *Introduction to Political Economy*, 132–156. Cambridge, Massachusetts: Economic Affairs Bureau.

11. Miller, Raymond. C. 2000. "Environmental Policy Implications of Clashing IPE Paradigms." Available from http://bss.sfsu.edu/ir/irjournal/ WinterSpring01/DrRaymondMiller.pdf; accessed July 3, 2005.

12. Dudley-Rowley, Marilyn. 2001. "The Globalization of Space." Pacific Sociological Association Annual Conference, San Francisco, California, 30 March. Another version, "The Globalization of Space in the 21st Century: Implications for the National Aeronautics and Space Administration," was submitted to the Price Waterhouse Endowment for The Business of Government, $15,000, January 2001–June 2001. Available from http://www.ops-alaska.com/GlobalizationOfSpace/PacSoc.htm; accessed July 3, 2005; and Dudley-Rowley, Marilyn, and Thomas Gangale. 2006 (September). "Sustainability Public Policy Challenges of Long-Duration Space Exploration." American Institute of Aeronautics and Astronautics San Jose, California.

13. Hardin, Garrett. 1968. "The Tragedy of the Commons." *Science*, 162:1243–1248. Available from http://dieoff.org/page95.htm; accessed February 22, 2006.

14. Frankel, Charles. 1956 *The Case for Modern Man.* Harper and Brothers, New York.

15. Hardin, "Tragedy of the Commons."

16. Ibid.

17. Dekanozov, R. V. 1980. "Juridicial Nature and Status of the Resources of the Moon and Other Celestial Bodies." *Proceedings, 23rd Colloquium on the Law of Outer Space.* American Institute of Aeronautics and Astronautics. 80-SL-03, 5–8.

18. Cooper, Lawrence A. 2003. "Encouraging Space Exploration Through a New Application of Space Property Rights." *Space Policy,* 19:111–118.

19. Smith, Adam. 1776. An *Inquiry Into the Nature and Causes of the Wealth of Nations.* Available from http://www.econlib.org/LIBRARY/Smith/smWN.html; accessed March 17, 2006.

20. Ricardo, David. 1817. *On the Principles of Political Economy and Taxation.* Available from http://internationalecon.com/v1.0/ch40/40c000.html; accessed March 16, 2006.

21. Wallerstein, Immanuel. 1999. "Patterns and Perspectives of the Capitalist World-Economy," In *International Relations Theory,* ed. Paul R. Viotti and Mark V. Kauppi, Boston: Allyn and Bacon, 369–376.

22. Spruyt, Hendrik. 1994. "Institutional selection in international relations" *International Organization,* 48:4.

23. Wikipedia. 2006. "Nation-State." Available from http://en.wikipedia.org/wiki/Nation-state; accessed March 18, 2006.

24. Gilpin, Robert. 2000. *The Challenge of Global Capitalism: The World Economy in the 21st Century.* Princeton: Princeton University Press, 22.

25. Steger, Manfred B. 2002. *Globalism.* Lanham, Maryland: Rowman & Littlefield, 28–29.

26. Ibid., 29.

27. Solana, Javier. 1998 (November 12). "Securing Peace in Europe." Symposium on the Political Relevance of the 1648 Peace of Westphalia, Münster. Available from http://www.nato.int/docu/speech/1998/s981112a.htm; accessed March 18, 2006.

28. Fischer, Joschka. 2000 (May 12). "From Confederacy to Federation—Thoughts on the Finality of European Integration." Speech Humboldt University, Berlin. Available from http://www.auswaertiges-amt.de/www/en/eu_politik/ausgabe_archiv?suche=1&archiv_id=1027&bereich_id=4&type_id=3; accessed March 19, 2006.

29. Berman, Yaniv. 2004 (April 1). "Al-Qa'ida: Islamic State Will Control the World." *akaJaneDoe.us.* Available from http://akajanedoe.us/islamicjihad12.html; accessed March 19, 2006.

30. Friedrichs, Jorg. 2001. "The Meaning of New Medievalism." *European Journal of International Relations,* 7,4:475–502

31. Whitney, Stewart B. 1984 (April 11–12). "Space Political Economy: Integrating Technology and Social Science for the 1990s." Third Annual Space Development Conference, San Francisco.

32. Seara Vázquez, Modesto. 1965. *Cosmic International Law.* Detroit: Wayne State University Press.

33. Gorove, Stephen. 1977. *Studies in Space Law: Its Challenges and Prospects.* Leyden: A. W. Sijthoff, 218.

34. Whitney, "Space Political Economy," 12–13.

35. Wasser, Alan. 1997 (March). "How to Restart a Space Race to the Moon and Mars." *Moon Miners' Manifesto, 103.* Available from http://www.asi.org/adb/ 06/09/03/02/103/space-race.html; accessed March 19, 2005.

36. White, Wayne N. 1998. "Real Property Rights in Outer Space." *Proceedings, 40th Colloquium on the Law of Outer Space.* American Institute of Aeronautics and Astronautics, 370. Available from http://www.spacefuture.com/archive/ real_property_rights_in_outer_space.shtml; accessed March 19, 2005.

37. McLuhan, Marshall. 1962. *The Gutenberg Galaxy: The Making of Typographic Man.* Toronto: University of Toronto Press.

38. Markoff, Marco G. 1970. "Space Resources and the Scope of the Prohibition in Article II of the 1967 Treaty." *Proceedings, 13th Colloquium on the Law of Outer Space.* American Institute of Aeronautics and Astronautics, 81–83.

39. Yandle, Bruce, Maya Vijayaraghavan, and Madhusudan Bhattarai. 2002 "The Environmental Kuznets Curve: A Primer," The Center for Free Market Environmentalism. Available from http://www.perc.org/perc.php?subsection= 9&id=688; accessed February 23, 2006.

40. Harbaugh, William T., Arik Levinson, and David Molloy Wilson. 2001 (February 27). "Reexamining the Empirical Evidence for an Environmental Kuznets Curve." Available from http://harbaugh.uoregon.edu/Papers/Environmental KuznetsCurve.pdf; accessed February 23, 2006.

41. McDougall, Walter A. 1985. *The Heavens and the Earth: A Political History of the Space Age.* New York: Basic Books.

42. Drexler, K. Eric, and Chris Peterson, with Gayle Pergamit. 1991. *Unbounding the Future: The Nanotechnology Revolution.* New York: Quill William Morrow.

10

The Cosmic Tumblers

WHY NOT

As with most large questions, that of why we are not moving out into space as aggressively as our technology would allow is one with many factors.

The general public is ignorant of space to a large degree. As a society, we have a poor understanding of our place in the solar system: the relative distances between Earth, the Moon, the sun, and the various planets and their moons; the relative sizes of these celestial bodies; the environmental conditions on them. In 2003, I observed firsthand how bogus satellite imagery of the February 1 destruction of *Columbia* and the August 14 electrical blackout of the northeastern United States and eastern Canada passed uncritically through the e-mail system among engineers at one of the nation's largest utility companies. One would have thought that such technically trained people would not have been so easily duped. Astrosociologist Jim Pass's distinction is apt: we have yet to become a true space-faring civilization, we are merely a space-capable civilization.[1]

The public generally supports the civil human space program, although it has little knowledge of what it is actually doing. And not only does the public not know what it is getting for its tax dollar, neither does it have any idea of what it is paying for. Polls show that only about 10 percent of the public correctly estimates that the National Aeronautics and Space Administration budget represents less than one percent of federal spending, whereas approximately 20 percent of the public believes that the NASA budget accounts for more than a quarter of federal expenditures.[2] This suggests that there would be much greater support for the civil space program if the public knew what a bargain they were getting, and

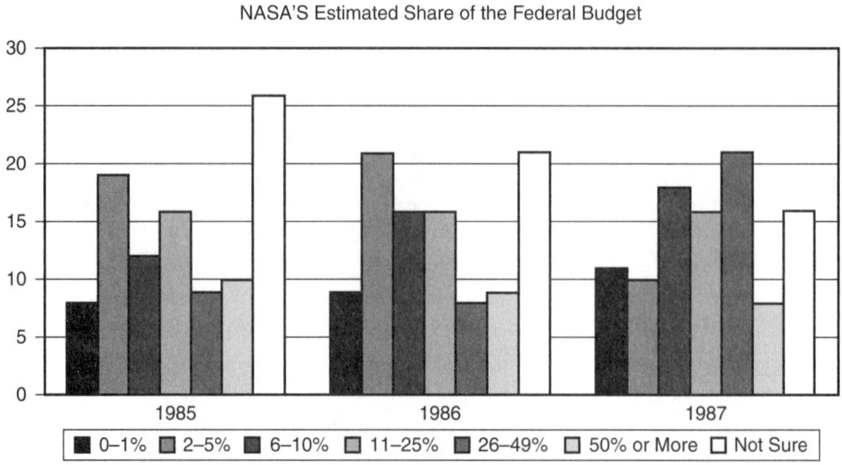

NASA'S Estimated Share of the Federal Budget

Figure 10.1: The Public's Estimate of NASA's Budget

Source: Roger D. Launius, National Air and Space Museum

might support spending levels several time higher than the current levels (compare Figures 9.1 and 10.1).

An activist citizenry has a positive role to play in promoting the "outward course of empire", in inculcating a general awareness that outer space is already an important sector of the global economy, and in ensuring that the importance of outer space will continue to grow.[3] This might have small effects on technoeconomy, educating entrepreneurs on the possibilities of outer space for profit-making enterprises. As threads in the fabric of the emerging civil society of outer space, they can be a valuable adjunct to the professional aerospace organizations, speaking as "the people," rather than as groups with obvious vested interests. Not having self-serving agendas accords them a certain specie in the corridors of power, an ability to be a faint, but uncorrupted voice in the wilderness. Operating in this venue, they can have an impact on technocratic decisions.

However, space enthusiast groups also have the defect of their virtue. They can be founded on certain unquestioned basic assumptions that should have been and ought to be questioned. As a case in point, the L-5 Society's misinterpretation of the Moon Agreement set it up to be co-opted into a corporate-led effort whose real target was the Law of the Sea Convention, whereas the torpedoing of the Moon Agreement was arguably against the interests of outer space development, the very mission of the L-5 Society. Space enthusiast organizations can be insular and polarized, preaching to their own choirs, either damning NASA, the major aerospace contractors, or the government-industrial complex in general as the Forces of Darkness, or singing hosannas to them as the Givers of Light. Often, the governance of these organizations is saddled with self-appointed presidents for life. These organizations tend to resemble personality cults, where the imams of outer space hawk their poorly vetted ideas year after year as they are superseded by events. For any of these reasons, ultimately, such groups are self-limiting and transient phenomena, reaching a maximum altitude then falling to Earth.[4] They soon squander whatever specie they might have had with policymakers at the outset. The founding conventions of these organizations may be well attended by optimistic professionals from government, industry, and academia, but after a few years, being associated with these groups can be as dangerous to one's career as dressing up like Marvin the Martian and chaining oneself to the White House fence. *That's an idea!*

The issue of property rights in outer space, and in particular the Space Settlement Prize Act, indicates that a considerable portion of the problem rests with the space advocacy community itself, but also the specialized media organs that provide the reportage for this community. Alan Wasser has been beating the drum on his patently bad idea for nearly 20 years, and in that time he has risen to high ranks in the space advocacy community and gained important endorsements for his idea. However, almost none of his adherents are cognizant of international law. Meanwhile, the many who are cognizant of international law and who object to his claims publish in scholarly venues but are largely ignored by the popular media. This allows Wasser to get away with labeling his critics as "dissenters" who just have another "opinion." This same popular media also dutifully reports the shady schemes of Dennis Hope and others, with little or no critical analysis.

In this environment, not only is it difficult to discern the well-intentioned but misguided idealists from the self-aggrandizing mountebanks, it is hard to tell these from the serious researchers. A case in point is the furor in early 2005 over a report of evidence of life on Mars:

> A pair of NASA scientists told a group of space officials at a private meeting here Sunday that they have found strong evidence that life may exist today on Mars, hidden away in caves and sustained by pockets of water.
>
> The scientists, Carol Stoker and Larry Lemke of NASA's Ames Research Center in Silicon Valley, told the group that they have submitted their findings to the journal Nature for publication in May, and their paper currently is being peer reviewed.
>
> What Stoker and Lemke have found, according to several attendees of the private meeting, is not direct proof of life on Mars, but methane signatures and other signs of possible biological activity remarkably similar to those recently discovered in caves here on Earth.[5]

You can't even have a casual conversation at a cocktail party these days without some enthusiastic lay-person misinterpreting you and blabbing to an irresponsible press that will print any hearsay.

"What I feel is that my privacy was violated," [Stoker] says. "From my point of view, what I had was a private conversation at a cocktail party. . . . I knew who all the people at the cocktail party were. I knew that none of them were reporters.

"There was a discussion about various things going on in the space program which we participated in. We probably were the only scientists in the room, and we were more privy to what was going on in the Mars community. We said some things that were going on that were not very different from what goes on in [public conferences]. . . ."

Stoker and Lemke declined all interview requests and NASA issued an unusual denial a few days later. . . .

Stoker now says that the damage was already done and that the conclusions attributed to her created initial skeptical impressions in the minds of her colleagues. . . .

As a result, she says, "you attract a reputation as somebody who isn't cautious, and isn't careful about what you say."

The "false alarm" did not just impact Stoker and Lemke, she says, but her colleagues in Spain as well.

"I think it damaged the reputation of the project," she says. "The impression I have is my collaborators in Spain were aghast, there were mumblings and grumblings from their scientific community that cast aspersions on the project."[6]

In aerospace engineering, the opposing forces on a vehicle in the direction of flight are thrust and drag. Those who believe in the proliferation of the human species throughout the Solar System as the imperative direction of flight, but who propound poorly vetted ideas and execute them in a poorly coordinated manner, not only discredit themselves and their associates, but also call into question the credibility of those who are engaged in serious endeavors. They believe they are contributing to the thrust of the vehicle, when actually they constitute a social force of drag on the mission. To employ another aerospace metaphor, the lack of an adequate guidance system in the space enthusiast community allows a lot of thrusting along different vectors that keeps the vehicle off course.

The citizen space community is small, fractured, and appears to lack a cohesive core of expertise. There is some pool of expertise, but it is diffused among a general population of enthusiasts, which makes it difficult to focus it on a productive purpose. In a positive development, the American Institute of Aeronautics and Astronautics, a professional organization that draws its membership from industry, government, and academia, has begun to increase its efforts in the policy arena, both through its annual Congressional Visit Day and through greater visibility of its Public Policy Committee. The citizen group ProSpace's annual "March Storm" on Capitol Hill is a noble effort; however, an informed voice that speaks both to civil society and in the corridors of power is sorely needed.

HOW

In writing about John F. Kennedy's decision to send men to the Moon in May 1961, historian Roger D. Launius argues that it is likely that Richard Nixon might have made the same decision had he won the election six months earlier; the flight of Yurij Gagarin aboard *Vostok 1* would still have been a Soviet triumph, and the Bay of Pigs invasion would probably have been an anti-Castro fiasco even with U.S. air support.[7] A History Channel series couches historical events in terms of *Man, Moment, Machine.* In the Moon decision, the machines were the rocket and the printed circuit board, and the moment was the Cold War, a month after the Bay of Pigs debacle; Launius doubts that in this moment, with these machines available, the man made much difference. The Apollo program was the child of the Cold War rivalry between the United States and the Soviet Union, and was specifically the child of the events of 1961. As Launius has described it, in May of that year the cosmic tumblers fell into place that enabled a decision to commit the nation "to achieving the goal, before this decade is out, of landing a man on the moon."[8] The decision was primarily the product of sociopolitical forces and technology.

If so, it may be that the next great decision regarding outer space is also not so much about the man, but about the moment and the machine. Speaking both as a rocket scientist and a social scientist, my estimate is that figuring out the machines is the easier of the two tasks. We put men on the Moon with 1960s technology, when computers were as big as your living room. Today, we keep two people in orbit around the Earth continuously, and operate robotic rovers on Mars for years at a time. If one counts Pluto as a Kuiper Belt Object, all of the planets in the Solar System have been explored; if one does not, there is a mission on its way to Pluto. The technologies are in hand to accomplish extraordinary things in outer space. All we have to do is integrate those technologies, build the infrastructure and the vehicles, train the ground and flight crews, and go. Oh, and one more thing: add money. And this is where we come down to the hardest questions. Who pays? Who benefits? What are the socioeconomic forces that will bring about the next surge into space? Sociologist Marilyn Dudley-Rowley observes:

> In truth, longer duration space missions have been possible for nearly three decades, as the Russians have so adequately shown during three generations of space stations. But, a long-duration space mission was not necessary during the early days of the space program when the primary goal was to beat the Russians to the moon. It is, however, necessary to almost anything else of value done in space, regardless of it being a robotic or manned mission; that value determined by the globalization of many aspects of populations, organizations, environments, and technologies.
>
> Now, it is a sociological fact that just because societies have on-the-shelf technologies to deploy, it does not follow that they will deploy them. Political, economic, and other forces, perceived and actual, have always delayed or denied entrée to some technologies in many societies throughout history. On the other hand, political, economic, and other forces are shaped by technology and technological feasibility.[9]

Technoeconomy is capable of remarkable accomplishments—when there are profits to be made. It takes time to nurture markets, and so progress is evolutionary. As explained in chapter nine, "Space Cowboys: Big Hat, No Cattle," the primary barrier is a drastic reduction in cost per mass to orbit. Advanced concepts for a single-stage-to-orbit (SSTO) launch vehicle proved to be too challenging. The first of these was Ronald Reagan's "Orient Express," also known as the X-30 and National Aerospace Plane (NASP). It began as a Defense Advanced Research Projects Agency (DARPA) project called Copper Canyon that ran from 1982 to 1985. Reagan took the wraps off in his 1986 "State of the Union" address, at which point the program became the X-30 NASP, funded by NASA and the United States Department of Defense.[10] Rockwell International won the competition as prime contractor.

> As a single-stage-to-orbit vehicle with a claimed turnaround time of as little as 24 hours, proponents of the Strategic Defense Initiative initially saw the X-30 leading the way to faster, cheaper access to low earth orbit, a critical aspect of lowering the cost of any space-based ballistic missile defense systems. However, as it became clear that the time required for the development of an operational capability would extend far beyond the time horizon envisioned for deployment of space-based anti-missile systems, the SDI program soon lost interest in the NASP effort. A similar disenchantment has emerged within the Air Force and NASA, as the high technical risk of the project has become increasingly clear. What has also become increasingly clear is that the claims made for NASP as a space launch vehicle are eerily reminiscent of the initial claims made for the Space Shuttle in the early

1970s. The assertions that NASP will have airplane-like operating characteristics, with lower costs and fast turnaround times on the ground, are
assumptions, rather than conclusions based on detailed analysis. . . .

A decision to undertake Phase 3 flight testing would have brought total
program costs up to as much as $17 billion. The target date for the first test
flight of the X-30 was pushed back to the 2000–2001 period, 11 years
behind schedule and 500 percent over budget. Many years and a further
$10 to $20 billion would have been required for the development of an
operational vehicle. . . .

Clearly, no single vehicle can serve commercial, civil space and military
masters at the same time. In spite of efforts to be all things to all people, the
NASP remained without a truly credible mission, and ultimately proponents were unable to save it from termination [in 1993]. . . .

The Hypersonic Systems Technology Program (HySTP), initiated in late
1994, was designed to transfer the accomplishments made in hypersonic
technologies by the National Aero-Space Plane (NASP) program into a technology development program.[11]

On January 27, 1995, the Air Force terminated its participation in
HySTP, ending the last gasp of NASP. Now, national security can be successfully invoked to justify an incredible range of technologies, so if it
could not pull off the NASP, it certainly was not for lack of trying.

On July 2, 1996, NASA selected Lockheed Martin to design, build, and
test the X-33, a subscale technology demonstrator for a next generation,
commercially operated space launch vehicle named VentureStar. The X-33
was to flight test a range of technologies needed for SSTO reusable launch
vehicles (RLVs). Construction of the prototype was about 85 percent complete when NASA canceled the program in 2001, after a long series of
technical problems, such as flight instability and excessive weight. NASA
had invested $912 million in the project before cancellation, and Lockheed
Martin had invested a further $357 million.

Based on the X-33 experience shared with NASA, Lockheed Martin hoped to
make the business case for a full-scale SSTO RLV, called VentureStar, that
would be developed and operated through commercial means. The intention
was that rather than operate space transport systems as it has with the Space
Shuttle, NASA would instead look to private industry to operate the reusable
launch vehicle and NASA would purchase launch services from the commercial launch provider. Thus, the X-33 was not only about honing space flight
technologies, but also about successfully demonstrating the technology
required to make a commercial reusable launch vehicle possible.[12]

Here we seem to have had the prefect partnership of a government
agency that wanted to get out of the launch operations business as a cost

center and a corporation that wanted to get into the business as a profit center; yet the technical challenges put their goals beyond the resources they were willing to risk. Certainly these sums are far beyond what the "new space entrepreneurs," the nimble and innovative startups, can command. Boosters for these new entities disparage the "cost plus" mentality of the major aerospace contractors that are used to the billion-dollar federal gravy train, and claim that the Burt Rutans and Jim Bensons of the world can do better for much less. These are only claims, some of which come from "companies" that are nothing more than a box of business cards and a West Hollywood apartment, perhaps even several of them in the same apartment. Rutan and Benson have remarkable accomplishments to their credit—on small projects. The scalability of the micro-entrepreneurial paradigm to giga-scale space development projects is unproven.

Technocracy has its ills, obviously. Writing in the 1980s, Walter McDougall characterized the Soviet Union as the world's first technocratic state; technocracy carried the Soviet Union only as far as 1991.[13] While stationed as an Air Force officer at a major aerospace contractor's facility at the height of the Cold War, I observed the high degree to which the contractor's organization mirrored that of its primary customer: the Department of Defense. However, that was before a decade and a half of post–Cold War mergers and downsizing. The surviving companies have been the more efficient ones, often diversifying into commercial lines of business. This is not to say that they cannot do better still, but it does suggest that when they must, they will.

The mini-entrepreneurs will certainly have a significant role in developing outer space along with the established aerospace companies. We need the best practices of both, and promoting an either-or dichotomy is misguided. The large picture is that developing outer space will take enormous sums of money and decades of effort, mostly on the taxpayers' dime, before there is enough there to sustain commercial operations. The next big push into space, taking us back to the Moon and further outward to Mars, is therefore likely to be the result of a technocratic decision, as was the Cold War race to the Moon. In ratifying the Moon Agreement, and in initiating and taking a leading role in the process of crafting a follow-on treaty to establish a governing regime for the solar system, the United States would send out a strong signal that it is ready to move forward in partnership with other states and with nongovernmental entities. The State Department and NASA should consider what they could do to support the development of the Regency of United Societies in Space as an incremental step to a more detailed legal regime defining and protecting property rights.

Outdated

WHEN

In the long run, the technocratic model might play out as a transnational project on a fairly relaxed timeline, in the absence of a major political problem to which a major, accelerated space effort can be seen in some light as a solution.[14] In this case, although NASA is now planning its *note* "return to the Moon as early as 2015 and no later than 2020," this could be stretched out, and the program might also be downscaled to the point that the path to sustained, profitable development of the Moon might be very lengthy, possibly not occurring until midcentury or later.[15] There is presently no timetable for Mars, whereas in 1969 NASA proposed a manned Mars mission for 1981, and in 1989, President George H. W. Bush proposed one before 2019. As of 2008, it appears that an expedition to Mars might occur in the 2030s, with sustained, profitable development of Mars occurring perhaps a century later. In short, if 1961 were 1492, then 2007 is only 1538; and there is a long way to go before the Pilgrims land at Rupes Plymouthensis.*

On the other hand, if China persists as an autocratic state as it climbs to the top of the economic heap, this enhances the possibility of rivalry between it and the "liberal pacific union" in a new space race.[16] The United States is likely to partner with other liberal democracies on major space projects, but will probably be reluctant to do so with China because of technology transfer concern, especially if China's geopolitical goals remain uncertain. Evidence in favor of this conclusion is that the United States invited post-Soviet Russia to partner with it, Europe, Japan, and Canada on the International Space Station; China was not invited, and it is making plans to assemble its own space station beginning in 2012. As a normative proposition, it would be nice to think of humankind journeying to the planets together; however, it is undeniable that competition spurs innovation, and so long as the international legal regime of outer space remains intact, and national sovereignty and private property rights are not asserted over territory, extending such a geopolitical rivalry into space might be fairly benign, as was the space race of the 1960s.

In a space race, it is possible that the return to the Moon could be accelerated by a couple of years, but since the Bush timetable appears adequate to get there ahead of China anyway, this is unlikely. Or is it? On September 17, 2007, NASA administrator Michael Griffin stated his belief "that China will be back on the Moon before we are."[17] Given an initial operational capability to perform lunar missions in 2015, and given that in 1969 NASA proposed to launch the first manned mission to Mars in 1981, a

---

* "Plymouth Rock" in Latin, the language of lunar and Martian geography.

race to Mars might result in a first expedition as early as 2027. An early decision to pursue a manned Mars program in parallel with the return to the Moon might pull this date forward to the early 2020s. There is no obvious reason why this decision would be made. But then, the 1961 technocratic Moon decision is only obvious in hindsight.

In his classic study of the governmental decision-making process, Graham T. Allison analyzes the 1962 Cuban Missile Crisis through the lenses of three paradigms:[18]

- The Rational Actor paradigm is the most economical paradigm of decision making, to use a double entendre. It is the most parsimonious of the three paradigms; it achieves this parsimony by employing many of the "rational choice" assumptions found in mainstream economics. It treats governments as monolithic, rational actors that have access to perfect information and select options that maximize value.
- The Organizational Process paradigm treats governments as constellations of constituent organizations, each of which has a sphere of responsibility and expertise, and each of which views the problem from its own perspective: how the problem threatens the organization's interests, and how the organization can provide the solution. These steps of characterizing the problem, weighing options, formulating the solution, and implementing it, are all performed according to the standard operating procedures to the maximum extent that suffices to achieve the organization's goals.
- The Governmental Politics paradigm adds another layer of nuance. At the top of each organization is its leader, each of whom has his individual set of strengths and weaknesses in terms of expertise, character, and values. The actors possess varying levels of power by virtue of the positions they occupy. Also, each has a varying degree of influence based on their personal relationships with other actors. These combinations of positions and relationships provide each actor certain specific action channels.

Allison develops the five plausible explanations for the Soviet Union's decision to deploy offensive SS-4 medium range ballistic missiles (MRBMs) and SS-5 intermediate range ballistic missiles (IRBMs) in Cuba based on the Rational Actor paradigm, and shows how each of them fail to explain all of the historical data. He then uses the Organizational Process paradigm to explain the anomalous historical facts that defy the Rational Actor paradigm, not the least of which are why the Soviets made no attempt to camouflage the missiles until after Kennedy announced their discovery, given that the success of the missile deployment hinged on the Soviets' ability to present Kennedy with a fait accompli, or why surface-to-air missile sites to defend the offensive missiles against almost certain U.S. air attacks were not prepared in advance. Analogously, we can see that the technocratic decision to develop the Space Shuttle had at least as much to do with dysfunctional organizational processes as it did with rational actors, to put it kindly.

Finally, Allison offers a fascinating view of how various considerations regarding the "missile gap" in the favor of the United States and the need to defend Cuba against future U.S. aggression played out against the political problems of particular Presidium members. Fidel Castro's purge of Cuban Communist Party leader Anibal Escalante created doubt as to how committed Castro was to the Communist Bloc, and that created a political problem for Mikhail Suslov, chief ideologue of the Communist Party of the Soviet Union (CPSU), and Boris Ponomarev, chief of the International Department in the CPSU Central Committee. The strategic gap translated as a political problem for defense minister Rodion Malinovskij. The failure of Nikita Khrushchev's 1961 gambit in Berlin was a political cloud he had to move out from under. The Soviet economy was expanding more slowly than expected, which was Alexei Kosygin's immediate concern, but this also meant that the heavy investment in intercontinental ballistic missiles (ICBMs) that would be necessary to close the gap with the United States would be difficult to sustain. In the end, these many internal political problems had one solution: deploy the MRBMs and IRBMs in Cuba.

Similarly, the timing of the next technocratic decision in outer space might be heavily influenced by political factors having little to do with outer space. Also, the acuteness of a political crisis would certainly drive the timetable of such a project. For instance, a new crisis in the War on Terrorism would be unlikely to precipitate an astropolitical technocratic response. Iran is only a nascent space launching state, and no other Islamic country is a space launching state. Outer space politics has even less relevance in the context of terrorist nonstate actors. On the other hand, China is a long-established launching state with an manned program, it is clearly interested in developing its space capabilities, and under its current form of government it is unlikely to partner with the United States. The near-simultaneity of a spectacular Chinese space achievement and a crisis over Taiwan is the sort of coupling of space politics and geopolitics that could trigger an astropolitical technocratic response.

Another consideration will be the degree to which a presidency can claim the project as its legacy. The Apollo landings on the Moon during the Nixon administration were unquestionably the legacy of John F. Kennedy and Lyndon B. Johnson, since the Moon decision came in the first months of the Kennedy administration, and was carried on by Johnson, who as

Senate majority leader had also led the effort to create NASA and the Mercury program in 1958. Although it was Richard Nixon who congratulated Neil Armstrong and Edwin Aldrin in July 1969, he was lukewarm at best to a program that was indelibly associated with his predecessors, and he funded only a small fraction of the missions that had been planned under the Apollo Applications Program to take maximum advantage of the technological and infrastructural capabilities developed during Apollo. The Apollo program was allowed to wither and die by 1975, whereas a more vigorous program could have provided the stepping stone to Mars expeditions in the early 1980s.

Thus, a flagship project in outer space, taking many years to complete, has its best chance of being launched early in a president's first term so that it is well on the way to completion by the end of the second term. A second possibility—perhaps a better one given the long programmatic timelines that span presidencies—is for an incoming administration to take an existing, low-key effort and transform it to the point that it is able to embrace it as its own. After all, Kennedy did not make his Moon decision in a technological vacuum; rather, NASA and its contractors had been studying concepts for a post-Mercury spacecraft and various mission options—including manned lunar landings—for a year before his May 1961 announcement. This family of concepts already had a name: Apollo.

When George W. Bush announced his Vision for Space Exploration in January 2004, it caused barely a political ripple, overshadowed as it was by the unfolding disaster in Iraq, and no mention of it was made during that year's presidential campaign. Three years later, the general public remained largely ignorant of the program to return humans to the Moon and to proceed to Mars expeditions, and it is likely to remain so through the end of the Bush presidency as the controversy over the occupation of Iraq continues to consume most of the oxygen in the national political forum. Thus the Constellation program may face a crisis in 2009–2010 as the Obama administration considers its options. On the one hand, there is the danger of the program being canceled outright; on the other hand, the Constellation program may acquire a new, more vigorous champion who will see the political value of placing the next telephone call from the Oval Office to the Moon.

HOW NOT

If there is a technocratic decision in reaction to geopolitical events arising out of a new Cold War with China, statecraft must keep any rivalry in outer space from getting out of hand. Again, the wisdom of the Outer

Space Treaty, in keeping large-scale national sovereignty out of space, is apparent. In the words of Kennedy:

> For space science, like nuclear science and all technology, has no conscience of its own. Whether it will become a force for good or ill depends on man, and only . . . [we can] decide whether this new ocean will be a sea of peace or a new terrifying theater of war. I do not say that we should or will go unprotected against the hostile misuse of space any more than we go unprotected against the hostile use of land or sea, but I do say that space can be explored and mastered without feeding the fires of war, without repeating the mistakes that man has made in extending his writ around this globe of ours.[19]

As noted in "Social Balance in Space" in chapter nine, the technoeconomy-technocracy dichotomy is not absolute; there is a push-pull relationship. Our walkabout the cosmos may be a series of complementary steps, the left foot of technocracy followed by the right foot of technoeconomy.

Furthermore, it is possible that a transnational project for the human exploration of the Moon and Mars might include not only nation-state partners, but corporate partners. Socializing some portion of the investment may incentivize private interests to put some of their own skin in the game. Nongovernmental organizations (NGOs), while commanding far smaller resources, might nevertheless find useful roles. As human space projects become larger, more complex, and more expensive, programmatically spanning several decades, the political-economic issues of funding continuity, and the organizational challenges of strategic partnerships between government agencies, private enterprises, and NGOs, all of different nationalities, become more challenging as well. There will be trade-offs.

For instance, in return for the multiplicity of funding sources and the stability this provides, there will be a loss of programmatic flexibility. This is not necessarily all bad. The U.S. Congress funds projects from year to year, and it can be feast or famine, depending on the political deals that are made in the corridors and the anterooms of the Capitol. Programs can slowed down or be knocked off track, then in the next fiscal year receive full funding. This is no way to run a development program. Every such program has an optimum funding curve that optimizes cost, schedule, and technical risk. Any deviation from that curve results in an increase of one, and usually more, of these three factors. If you speed up the project, it costs money. If you slow down the project, it costs money. Such deviations are the norm, not the exception, in federally funded projects.

At the "New Trends in Astrodynamics II" Symposium held at Princeton University in June 2005, Hayden Planetarium director Neil deGrasse Tyson, who served on the President's Commission on Implementation of United States Space Exploration Policy (Aldridge Commission), which issued its report in 2004, gave a presentation on George W. Bush's "New Vision for the Space Exploration Program" (not published in the symposium's proceedings).[20] As I recall, he discussed how the administration's goals regarding the human exploration of the Moon and Mars would be accomplished with modest, sustainable funding. That sounds fine as far as it goes, but if funding should become an issue, that is, when the program hits some technical snag that is more difficult than anticipated and therefore more expensive to solve, Tyson responded, "We'll just stretch it out."

This is great grist for the space libertarian mill, for it calls into question the credibility of the government's commitment to human space exploration. Any technical manager understands that one budgets a certain percentage of a project's schedule and money as "management reserve." This is the insurance policy to handle the "unknown unknowns," the unanticipated technical challenges. In the long run, management reserve saves money, because when it has to be spent, it keeps the project on its optimum curve.

"Stretching it out" costs money. The International Space Station (ISS) is a splendid example of this. The project began in 1984, and was targeted to be completed in 1994 at a cost of $14.5 billion. It is not yet completed, is considerably down-scoped from its original design (despite having absorbed modules originally designed for the Soviet *Mir 2*), and is now expected to cost over $100 billion.[21] This is just the cost overrun to the U.S. taxpayer. In addition, the schedule slips on the U.S. side have caused cost overruns in the Russian, European, Japanese, and Canadian components of the ISS. The European Space Agency's (ESA) history of cooperation with NASA has been particularly disappointing. ESA developed the Spacelab for the Space Shuttle as a modular, versatile, multi-mission series of payloads capable of many reflights. Because the Space Shuttle proved incapable of operating the high number of missions per year that was envisioned in the original traffic models, and Spacelab missions were a low priority for the U.S. government compared to the Shuttle-specific designs of certain national security spacecraft, NASA flew only a handful of Spacelab missions. As a result, ESA invested a lot of money and received very little return. If they know what is good for them, national governmental and transnational corporate partners will hold the U.S. government's feet to the fire on any future space projects by insisting on serious and enforceable contractual consequences to material breaches. In a multiple funding source project, all parties should commit to full funding of their assigned portions from one major programmatic milestone to the next, as defined in a common project management document such as MIL-STD-1521B: system design reviews, preliminary design reviews, critical design reviews, etc.[22] At each of these decision points, the parties should reach a consensus as to whether the project should continue to the next major milestone with full funding.

"Stretching it out" and inadequate cost and schedule flexibility can even end up costing lives:

> The Lady Franklin Bay Expedition in the Eastern [Canadian] Arctic [1881–1884] was organized in response to an international polar science project. Officially mandated, it was shabbily put in motion because of Congressional and military mis-coordination. When the expedition got in trouble owing to the bumbling of the pick-up ships under the command of the U.S. Navy, the government simply abandoned 25 healthy expeditioners. As a result, only seven men survived, and Commander Adolphus Greeley has gone down in popular history with the "bum rap" of being an incompetent leader. Digging into the facts of the expedition, however, one finds that it was the efforts of Greeley's wife, who through much networking and private means, got a bounty imposed among international naval, whaling, and trading vessels for the rescue of the party. Without her effort, there would have been no survivors.[23]

When the scheduled pickup failed, Congress saw no point in funding a second attempt, since the expedition members were undoubtedly dead. But, in the Space Age, the government will never risk human lives because of cost and schedule concerns, right? The decision to launch OV-099 *Challenger* mission STS-51L on a frosty January morning was driven to a high degree by cost and schedule pressures.

Another concern regarding a technocratic decision for the next big push into space is that it must serve the right mix of interests and goals. Within some factions of governments, there is merely the desire to achieve a goal for national prestige, a "flags and footprints" project that results in a few expeditions and then terminates, as the Apollo program did, leaving no in situ infrastructure. Other factions in the bureaucracy will be more attuned to corporate interests and civic organizations that would rather see a sustained program leading to the profitable development of outer space resources. "Different-colored money" flowing into the project coffer from national governments, private companies, and civic organizations better ensures a sustainable project. Of course, a project that is meant to serve multiple users can get very fouled up. The Space Shuttle was envisioned as a "space-going TFX"* that would somehow

---

* TFX (Tactical Fighter Experimental) was the programmatic designation for the F-111 aircraft, which was designed in the early 1960s to be not just a fighter for the Tactical Air Command but a nuclear bomber for the Strategic Air Command, and not just for the Air Force but for the Navy as well; however, when production began, the Navy refused to buy any F-111s.

be able to support whatever the United States might want to do in space, whenever it made up its mind. It would serve NASA's interests in sustained human spaceflight and space science, the Defense Department's and Central Intelligence Agency's interest in routine access to space for national security assets, and the commercial satellite industry's interest in inexpensive launch services. The Space Shuttle never did any of these well.

WHY

Nicholas van Rijn and Jim Benson aside, profit is not the only reason to go into space; money is not the only measure of value. National prestige can sustain a certain level of effort for nonmilitary programs over a period of decades; in the United States, that level has been about one percent of the federal budget. Military programs to project national power can command several percent of the federal budget. A politically motivated display of national technological power may cause a technocratic spike in space activity. If there were no money to be made in space, if there were no national security strategies in space, it is true that many would shrug and say, "What use is it?" But it is also true that some would understand that this tiny Earth of ours is subject to forces far above its atmosphere. Earth is in outer space, therefore we live in outer space. Regardless of whether we take a moment from our mundane existence to reflect on that fact, it is nevertheless fact, a fact that the dinosaurs could not comprehend as the Cretaceous Period went out with a bang 65 million years ago.

Developing a spacefaring culture is a matter of survival, not just to gain the ability to detect and deflect asteroids and comets on a collision course for Earth, but to escape the resource constraints of out limited planet. There are perhaps 30 years of petroleum left; let's hope that by the time it runs out, commercial fusion power (perhaps fueled by the Moon's helium-3) or solar power satellites are up and running, waiting for the baton to be passed. If not, the Great Machine on which Earth's billions depend could shudder to a halt. Given the global population of 800 million that a rudimentary industrial economy supported a couple of centuries ago, collapse of the Great Machine could mean death for 90 percent of Earth's population toward the middle of the twenty-first century, death by starvation, opportunistic diseases, and resource wars. I am not saying that the end of the world is nigh, nor am I saying that the only path to avoiding the Apocalypse leads into space, but I am saying that Earth-based solutions to the end of oil may not be entirely adequate. In space, there is the possibility of developing other options, and it would be wise to have them available should we need to exercise them. While the business case for space tourism pioneering cheap and large-scale spacelift capability is questionable, the case for keeping the Great Machine running is obvious.

But even if spacefaring were not a matter of survival, we would still need to go into space, because the answers to our very existence are out there, and curiosity is one of the strongest of human traits. Where did we come from? How did we get here? Who else is out there?

For more than a century, these questions have been part of the public discourse about outer space. Inspired by the speculations of Camille Flammarion,[24] Percy Greg,[25] Percival Lowell,[26] Kurd Lasswitz,[27] and H. G. Wells,[28] the optimism and confidence of Victorian civilization took it almost as axiomatic that it was on the verge of making "Contact" with a sentient species on Mars. At the turn of the twentieth century, the *New York Times* reported Nikola Tesla's plans to send radio waves to Mars and communicate with its inhabitants.[29] As we better acquainted ourselves with Mars in the scientific sense in the course of the twentieth century, there came, as Wells wrote, "the great disillusionment." We came to realize that in terms of sentient species, we are alone in the solar system. Yet a faded echo of Lowellian Mars remains. We cling to the hope of a neighboring planet that harbors, if not canals and an advanced civilization, at least some primitive forms of life. If Mars contains even nanobacteria—or indisputable evidence of past life of the simplest forms—this will profoundly change our conception of our place in the universe. If there is—or was—another Genesis here in our own solar system, then life must be common throughout the universe, and "Contact" with another civilization is therefore inevitable.

Do we need to send humans to Mars to discover this? No, not necessarily. It is possible that robotic missions to Mars could make such a startling discovery. But machines alone are not as capable as humans and machines working together in situ. So, if robots do not find life on Mars, the question remains open, even if just a crack. Eventually, we humans must go to Mars ourselves to definitively satisfy our curiosity.

If Mars is dead now, but was once alive, understanding how Mars died may give us a crucial understanding of how close we are coming to killing the Earth. Knowing Mars will allow us to calibrate our knowledge of Earth. As forbidding an environment as we have come to know Mars to be in the past few decades, it is nevertheless the most Earth-like planet in the solar system, the most readily accessible from Earth, and given sufficient technology and infrastructure, it will be able to support human life.

Is it worth spending tens of billions, possibly hundreds of billions of dollars, to send humans to Mars? In considering this prospective question, it is useful to ask a retrospective one: was it worth it to send humans to the Moon?

There are certain indelible images of the age of photography: Battleship Row in Pearl Harbor on December 7, 1941; the Zapruder film of Dealey Plaza on November 22, 1963; the twin towers of the World Trade Center on September 11, 2001. These not only capture specific events, but also define the specific locales and eras in which they occurred. But the images of the Earth that we brought back from the Moon are timeless and universal, because they are the first images of all of us. Ever since then, because of those images, we have looked at ourselves, each other, and the Earth in a new way. The image of the full Earth brought back by the last crew to return from the Moon is an enduring icon of environmental responsibility and human unity. Was it worth going to the Moon to bring back even one of those photographs of Earth? I believe that it was.

Figure 10.2: The Whole Earth

Source: NASA (From *Apollo 17* during its return from the Moon)

The most important thing that we discovered on the Moon was part of ourselves. In the few hours that a few of us spent on the Moon between 1969 and 1972, we became better Earthlings. As the poet Archibald MacLeish wrote, we were "riders on the Earth together."[30] We realized that we were our brother's keeper, and we remembered that God had appointed us stewards of the Earth. And yet, a third of a century later, we must reflect on how pitifully less we have done with that revelation than we should have. It is high time that we journeyed outward to that distant perspective, to see again how close we really are to each other, and to relearn those lessons that have faded with the passing of a generation. There are new lessons to be learned on Mars. There are new poems waiting for us on Mars.[31]

And then, it is on to Europa, on another quest for life.

Civilizations have risen, fallen, and in time others have risen in their place, but this time the stakes are greater. If, for some reason, our technological civilization should collapse, either because of nuclear war, pandemic, climate change, cosmic impact, or resource depletion, we can never pass this way again. No previous culture has massively consumed nonrenewable resources as ours has. Each decade that passes, we must dig deeper and drill farther to extract the materials that fuel the Great Machine. The advance of technology continually extends our reach for these resources, but these advanced methods would be far beyond the grasp of a postapocalyptic agrarian culture trying to make another go of it. What we think of as nonrenewable resources actually are renewable of course—on a geologic time scale. Left to itself, the Earth would again form subterranean pools of petroleum. Another Industrial Revolution might be possible on this planet, but only for a species as far removed from us in the future as the trilobites are in our past. Our civilization has the one and only chance the human race will ever have to reach beyond this planet and establish itself elsewhere in the universe. If we miss this opportunity, our species will be bound to the Earth until we become extinct.

If, on the other hand, we survive the various threats to the progress of technological civilization, we will see a branching of the human timeline. Humans will go to live and work indefinitely on orbiting space platforms, in lunar settlements, on Mars, and then out to the planet-sized moons of the gas giants. The process of inhabiting and thriving in ever more extreme environments is the natural extension of the coldward course of progress, the process by which humans left their tropical home-of-origin and ventured into the temperate and polar zones. The experience the solar system explorers, pioneers, and settlers will gain will pave the way to the stars—and beyond. As visionary scientist Carl Sagan pointed out, this gets the human eggs out of the single basket in terms of any sort of catastrophic mass extinction event.[32] It also gets our eggs out of the basket in terms of the natural processes of passive extinction, where we lose so much genetic vigor that we can no longer cope with our constantly changing single planetary environment. Because of the distances involved, not to mention the effects of wholly new planetary environments, in journeying outward we set in motion new speciation and differentiation of the *Homo sapiens sapiens* line. For our species to survive, we must diffuse into the cosmos. We must engage the grand environment, and who can say for how long our window of opportunity will remain open?[33]

On Earth, the best we can look forward to is a future of husbanding limited resources, some renewable, others inexorably dwindling. Although reaching beyond Earth in a sustained effort does not lessen our duty to manage Earth responsibly, it opens possibilities as yet unfathomable. The promise of space, although easy to oversell and challenging to fulfill, is impossible to abandon.

NOTES

1. Pass, Jim. 2004. "Space: Sociology's Forsaken Frontier." Available from http://www.astrosociology.com/Library/PDF/submissions/Space_Sociology%27s%20Forsaken%20Frontier.pdf, accessed March 25, 2009.

2. Launius, Roger D. 2000. "Project Apollo in American Memory and Myth." In *Proceedings of Space 2000*, eds. Stewart W. Johnson, Koon Meng Chua, Rodney G. Galloway, and Philip I. Richter. Reston, Virginia: American Society of Civil Engineers, 1–13.

3. Dudley-Rowley, Marilyn. 1999. "The Outward Course of Empire: The Hard, Cold Lessons From American Involvement in the Terrestrial Polar Regions." Mars Society. MAR 98-009. Founding Convention of the Mars Society. Boulder, Colorado, August 13, 1998. In *Proceedings of the Founding Convention of the Mars Society*, Volume I, eds. Robert M. Zubrin and Maggie Zubrin. San Diego, California: Univelt. Available from http://www.ops-alaska.com/mars/OutwardCourseOfEmpire.htm; accessed March 19, 2006.

4. Michaud, Michael A. G. 1986. *Reaching for the High Frontier: The American Pro-Space Movement, 1972–1984.* New York: Praeger.

5. Berger, Brian. 2005 (February 16). "Exclusive: NASA Researchers Claim Evidence of Present Life on Mars." Space.com. Available from http://www.space.com/scienceastronomy/mars_life_050216.html; accessed March 23, 2006.

6. Oberg, James. 2005 (March 22). "Scientist at Center of Mars Flap Speaks Out." MSNBC.com. Available from http://www.marsnews.com/newswire/life_on_mars/; accessed March 23, 2006.

7. Launius, "Project Apollo."

8. Kennedy, John F. 1961 (May 25). "Special Message to the Congress on Urgent National Needs." Available from http://www.jfklibrary.org/j052561.htm; accessed January 7, 2006.

9. Dudley-Rowley, Marilyn. 2001 (March 30). "The Globalization of Space." Pacific Sociological Association Annual Conference, San Francisco, California. Another version, "The Globalization of Space in the 21st Century: Implications for the National Aeronautics and Space Administration," was submitted to the Price Waterhouse Endowment for The Business of Government, $15,000, January 2001–June 2001. Available from http://www.ops-alaska.com/GlobalizationOfSpace/PacSoc.htm; accessed July 3, 2005.

10. Reagan, Ronald. 1986. "State of the Union Address." Available from http://janda.org/politxts/state of union addresses/1981-1988 Reagan/RWR86.html; accessed October 23, 2008.

11. Pike, John. 1997. "X-30 National Aerospace Plane (NASP)." Federation of American Scientists. Available from http://www.fas.org/irp/mystery/nasp.htm; accessed March 20, 2006.

12. Wikipedia. 2006. "Lockheed Martin X-33." Available from http://en.wikipedia.org/wiki/Lockheed_Martin_X-33; accessed March 20, 2006.

13. McDougall, Walter A. 1985. *The Heavens and the Earth: A Political History of the Space Age.* New York: Basic Books.

14. Dudley-Rowley, Marilyn, and Thomas Gangale. 2006 (September). "Sustainability Public Policy Challenges of Long-Duration Space Exploration." American Institute of Aeronautics and Space. San Jose, California.

15. White House Office of the Press Secretary. 2004 (January 14). "President Bush Announces New Vision for Space Exploration Program." Available from http://www.whitehouse.gov/news/releases/2004/01/20040114-1.html; accessed March 20, 2006.

16. Doyle, Michael W. 1999. "Preserving and Expanding the Liberal Pacific Union." In *International Order and the Future of World Politics,* eds. T. V. Paul and John A. Hall. Cambridge: Cambridge University Press.

17. Morring, Frank. 2007 (September 18). "China To Explore Moon Sooner: Griffin." *Aviation Week.* Available from http://www.aviationweek.com/aw/generic/story.jsp?id=news/china091807.xml&headline=China%20To%20Explore%20Moon%20Sooner:%20Griffin&channel=space; accessed October 1, 2007.

18. Allison, Graham T. 1971. *Essence of Decision: Explaining the Cuban Missile Crisis.* Boston: Little, Brown.

19. Kennedy, John F. 1962. "Address at Rice University on the Nation's Space Effort." September 12. Available from http://www.jfklibrary.org/j091262.htm; accessed December 30, 2005.

20. Aldridge, Edward C., Carlton S. Fiorina, Michael P. Jackson, Laurie A. Leshin, Lester L. Lyles, Paul D. Spudis, Neil deGrasse Tyson, Robert S. Walker,

and Maria T. Zuber. 2004. *A Journey to Inspire, Innovate, and Discover: Report of the President's Commission on Implementation of United States Space Exploration Policy.* Available from http://www.nasa.gov/pdf/60736main_M2M_report_small.pdf; accessed March 18, 2005.

21. United States General Accounting Office. 1998 (June 24). "U.S. Life-Cycle Funding Requirements." Available from http://www.gao.gov/archive/1998/ns98212t.pdf; accessed November 7, 2008.

22. United States Air Force. 1985 (June 4). MIL-STD-1521B (USAF), "Military Standard: Technical Reviews and Audits for Systems, Equipments, and Computer Software." Available from http://sparc.airtime.co.uk/users/wysywig/1521b.htm; accessed March 21, 2006.

23. Dudley-Rowley, "The Outward Course."

24. Flammarion, Camile. 1862. *La pluralité des mondes habités; and* Flammarion, Camile. 1893. *La planète Mars et ses conditions d'habitabilité.*

25. Greg, Percy. 1880. *Across the Zodiac: The Story of a Wrecked Record.* London: Trübner & Co.

26. Lowell, Percival. 1895. *Mars.* Boston: Houghton-Mifflin; Lowell, Percival. 1906. *Mars and Its Canals.* New York: Macmillan; and Lowell, Percival. 1908. *Mars as the Abode of Life.* New York: Macmillan.

27. Lasswitz, Kurd. 1897. *Auf zwei Planeten.* Leipzig: Verlag B. Elischer Nachfolger.

28. Wells, H. G. 1898. *The War of the Worlds.* London: Heinemann.

29. *New York Times.* 1905 (January 15). "Interplanetary Telephone."

30. MacLeish, Archibald. 1968 (December 25). "Riders on Earth Together, Brothers in Eternal Cold." *New York Times.* Available from http://cecelia.physics.indiana.edu/life/moon/Apollo8/122568sci-nasa-macleish.html; accessed October 23, 2008.

31. Gangale, Thomas. 2005a. "Why Should We Send Humans to Mars?" In *Moving Along: Far Ahead,* Volume 4 of the "Tackling Tomorrow Today" textbook series, ed. Arthur B. Shostak. Philadelphia, Pennsylvania: Chelsea House Publishers. Available from http://www.ops-alaska.com/mars/Why_Humans_to_Mars.htm; accessed March 21, 2006.

32. Sagan, Carl. 1995. *Pale Blue Dot: A Vision of the Human Future in Space.* New York: Random House.

33. Dudley-Rowley, "The Outward Course."

Appendix 1

The International Cooperation Resolution

Resolution on International Cooperation in the Peaceful Uses of Outer Space

A

The General Assembly,

Recognizing the common interest of mankind in furthering the peaceful uses of outer space and the urgent need to strengthen international co-operation in this important field,

Believing that the exploration and use of outer space should be only for the betterment of mankind and to the benefit of States irrespective of the stage of their economic or scientific development,

1. *Commends* to States for their guidance in the exploration and use of outer space the following principles:

    (a) International law, including the Chapter of the United Nations, applies to outer space and celestial bodies;

    (b) Outer space and celestial bodies are free for exploration and use by all States in conformity with international law and are not subject to national appropriation;

2. *Invites* the Committee on the Peaceful Uses of Outer Space to study and report on the legal problems which may arise from the exploration and use of outer space.

1085th plenary meeting,
20 December 1961

B

The General Assembly,

Believing that the United Nations should provide a focal point for international co-operation in the peaceful exploration and use of outer space,

1. *Calls upon* States launching objects into orbit or beyond to furnish information promptly to the Committee on the Peaceful Uses of Outer Space, through the Secretary-General, for the registration of launchings;

2. *Requests* the Secretary-General to maintain a public registry of the information furnished in accordance with paragraph 1 above;

3. *Requests* the Committee on the Peaceful Uses of Outer Space, in co-operation with the Secretary-General and making full use of the functions and resources of the Secretariat;

   (a) To maintain close contact with governmental and non-govermental organizations concerned with outer space matters;

   (b) To provide for the exchange of such information relating to outer space activities as Governments may supply on a voluntary basis, supplementing but not duplicating existing technical and scientific exchanges;

   (c) To assist in the study of measures for the promotion of international co-operation in outer space activities;

4. *Further requests* the Committee on the Peaceful Uses of Outer Space to report to the General Assembly on the arrangements undertaken for the performance of those functions and on such developments relating to the peaceful uses of outer space as it considers significant.

1085th plenary meeting,
20 December 1961

C

The General Assembly,

Noting with gratification the marked progress for meteorological science and technology opened up by the advances in outer space,

Convinced of the world-wide benefits to be derived from international co-operation in weather research and analysis,

1. *Recommends* to all Member States and to the World Meteorological Organization and other appropriate specialized agencies the early and

comprehensive study, in the light of developments in outer space, of measures;

(a) To advance the state of atmospheric science and technology so as to provide greater knowledge of basic physical forces affecting climate and the possibility of large-scale weather modification;

(b) To develop existing weather forecasting capabilities and to help Member States make effective use of such capabilities through regional meteorological centres;

2. *Requests* the World Meteorological Organization, consulting as appropriate with the United Nations Educational, Scientific and Cultural Organization and other specialized agencies and governmental and non-governmental organizations, such as the International Council of Scientific Unions, to submit a report to the Governments of its Member States and to the Economic and Social Council at its thirty-fourth session regarding appropriate organizational and financial arrangements to achieve those ends, with a view to their further consideration by the General Assembly at its seventeenth session;

3. *Requests* the Committee on the Peaceful Uses of Outer Space, as it deems appropriate, to review that report and submit its comments and recommendations to the Economic and Social Council and to the General Assembly.

1085th plenary meeting,
20 December 1961

D

The General Assembly,

Believing that communication by means of satellites should be available to the nations of the world as soon as practicable on a global and non-dicriminatory basis;

Convinced of the need to prepare the way for the establishment of effective operational satellite communication,

1. *Notes with satisfaction* that the International Telecommunication Union plans to call a special conference in 1963 to make allocations of radio frequency bands for outer space activities;

2. *Recommends* that the International Telecommunication Union consider at that conference those aspects of space communication in which international co-operation will be required;

3. *Notes* the potential importance of communication satellites for use by the United Nations and its principal organs and specialized agencies for both operational and informational requirements;

4. *Invites* the Special Fund and the Expanded Programme of Technical Assistance, in consultation with the International Telecommunication Union, to give sympathetic consideration to requests from Member States for technical and other assistance for the survey of their communication needs and for the development of their domestic communication facilities, so that they may make effective use of space communication;

5. *Requests* the International Telecommunication Union, consulting as appropriate with Member States, the United Nations Educational, Scientific and Cultural Organization and other specialized agencies and governmental and non-governmental organizations, such as the Committee on Space Research of the International Council of Scientific Unions, to submit a report on the implementation of these proposals to the Economic and Social Council at its thirty-fourth session and to the General Assembly at its seventeenth session;

6. *Requests* the Committee on the Peaceful Uses of Outer Space, as it deems appropriate, to review that report and submit its comments and recommendations to the Economic and Social Council and to the General Assembly.

1085th plenary meeting,
20 December 1961

E

The General Assembly,

Recalling its resolution 1472 (XIV) of 12 December 1959,

Noting that the terms of office of the members of the Committee on the Peaceful Uses of Outer Space expire at the end of 1961,

Noting the report of the Committee on the Peaceful Uses of Outer Space,

1. *Decides* to continue the membership of the Committee on the Peaceful Uses of Outer Space as set forth in General Assembly resolution 1472 (XIV) and to add Chad, Mongolia, Morocco and Sierra Leone to its membership in recognition of the increased membership of the United Nations since the Committee was established;

2. *Requests* the Committee on the Peaceful Uses of Outer Space to meet not later than 31 March 1962 to carry out its mandate as contained in General Assembly resolution 1472 (XIV), to review the activities provided for in resolutions A,B, C and D above and to make such reports as it may consider appropriate.

1085th plenary meeting,
20 December 1961

Appendix 2

The Declaration of Legal Principles of Outer Space

Declaration of Legal Principles Governing the Activities of
States in the Exploration and Uses of Outer Space

The General Assembly,

Inspired by the great prospects opening up before mankind as a result of man's entry into outer space,

Recognizing the common interest of all mankind in the progress of the exploration and use of outer space for peaceful purposes,

Believing that the exploration and use of outer space should be carried on for the betterment of mankind and for the benefit of States irrespective of their degree of economic or scientific development,

Desiring to contribute to broad international co-operation in the scientific as well as in the legal aspects of exploration and use of outer space for peaceful purposes,

Believing that such co-operation will contribute to the development of mutual understanding and to the strengthening of friendly relations between nations and peoples,

Recalling its resolution 110 (II) of 3 November 1947, which condemned propaganda designed or likely to provoke or encourage any threat to the peace, breach of the peace, or act of aggression, and considering that the aforementioned resolution is applicable to outer space,

Taking into consideration its resolutions 1721 (XVI) of 20 December 1961 and 1802 (XVII) of 14 December 1962, adopted unanimously by the States Members of the United Nations,

Solemnly declares that in the exploration and use of outer space States should be guided by the following principles:

1. The exploration and use of outer space shall be carried on for the benefit and in the interests of all mankind.

2. Outer space and celestial bodies are free for exploration and use by all States on a basis of equality and in accordance with international law.

3. Outer space and celestial bodies are not subject to national appropriation by claim of sovereignty, by means of use or occupation, or by any other means.

4. The activities of States in the exploration and use of outer space shall be carried on in accordance with international law, including the Charter of the United Nations, in the interest of maintaining international peace and security and promoting international co-operation and understanding.

5. States bear international responsibility for national activities in outer space, whether carried on by governmental agencies or by non-governmental entities, and for assuring that national activities are carried on in conformity with the principles set forth in the present Declaration. The activities of non-governmental entities in outer space shall require authorization and continuing supervision by the State concerned. When activities are carried on in outer space by an international organization, responsibility for compliance with the principles set forth in this Declaration shall be borne by the international organization and by the States participating in it.

6. In the exploration and use of outer space, States shall be guided by the principle of co- operation and mutual assistance and shall conduct all their activities in outer space with due regard for the corresponding interests of other States. If a State has reason to believe that an outer space activity or experiment planned by it or its nationals would cause potentially harmful interference with activities of other States in the peaceful exploration and use of outer space, it shall undertake appropriate international consultations before proceeding with any such activity or experiment. A State which has reason to believe that an outer space activity or experiment planned by another State would cause potentially harmful interference with activities in the peaceful exploration and use of outer space may request consultation concerning the activity or experiment.

7. The State on whose registry an object launched into outer space is carried shall retain jurisdiction and control over such object, and any

personnel thereon, while in outer space. Ownership of objects launched into outer space, and of their component parts, is not affected by their passage through outer space or by their return to the earth. Such objects or component parts found beyond the limits of the State of registry shall be returned to that State, which shall furnish identifying data upon request prior to return.

8. Each State which launches or procures the launching of an object into outer space, and each State from whose territory or facility an object is launched, is internationally liable for damage to a foreign State or to its natural or juridical persons by such object or its component parts on the earth, in air space, or in outer space.

9. States shall regard astronauts as envoys of mankind in outer space, and shall render to them all possible assistance in the event of accident, distress, or emergency landing on the territory of a foreign State or on the high seas. Astronauts who make such a landing shall be safely and promptly returned to the State of registry of their space vehicle.

Appendix 3

The Outer Space Treaty

Treaty on Principles Governing the Activities of States in the
Exploration and Use of Outer Space, including the
Moon and Other Celestial Bodies

Signed at Washington, London, Moscow, January 27, 1967

Ratification advised by U.S. Senate April 25, 1967

Ratified by U.S. President May 24, 1967

U.S. ratification deposited at Washington, London, and Moscow October 10, 1967

Proclaimed by U.S. President October 10, 1967

Entered into force October 10, 1967

The States Parties to this Treaty,

Inspired by the great prospects opening up before mankind as a result of mans entry into outer space,

Recognizing the common interest of all mankind in the progress of the exploration and use of outer space for peaceful purposes,

Believing that the exploration and use of outer space should be carried on for the benefit of all peoples irrespective of the degree of their economic or scientific development,

Desiring to contribute to broad international co-operation in the scientific as well as the legal aspects of the exploration and use of outer space for peaceful purposes,

Believing that such co-operation will contribute to the development of mutual understanding and to the strengthening of friendly relations between States and peoples,

Recalling resolution 1962 (XVIII), entitled "Declaration of Legal Principles Governing the Activities of States in the Exploration and Use of Outer Space," which was adopted unanimously by the United Nations General Assembly on 13 December 1963,

Recalling resolution 1884 (XVIII), calling upon States to refrain from placing in orbit around the Earth any objects carrying nuclear weapons or any other kinds of weapons of mass destruction or from installing such weapons on celestial bodies, which was adopted unanimously by the United Nations General Assembly on 17 October 1963,

Taking account of United Nations General Assembly resolution 110 (II) of 3 November 1947, which condemned propaganda designed or likely to provoke or encourage any threat to the peace, breach of the peace or act of aggression, and considering that the aforementioned resolution is applicable to outer space,

Convinced that a Treaty on Principles Governing the Activities of States in the Exploration and Use of Outer Space, including the Moon and Other Celestial Bodies, will further the Purposes and Principles of the Charter of the United Nations,

Have agreed on the following:

Article I

The exploration and use of outer space, including the moon and other celestial bodies, shall be carried out for the benefit and in the interests of all countries, irrespective of their degree of economic or scientific development, and shall be the province of all mankind.

Outer space, including the moon and other celestial bodies, shall be free for exploration and use by all States without discrimination of any kind, on a basis of equality and in accordance with international law, and there shall be free access to all areas of celestial bodies.

There shall be freedom of scientific investigation in outer space, including the moon and other celestial bodies, and States shall facilitate and encourage international co-operation in such investigation.

Article II

Outer space, including the moon and other celestial bodies, is not subject to national appropriation by claim of sovereignty, by means of use or occupation, or by any other means.

Article III

States Parties to the Treaty shall carry on activities in the exploration and use of outer space, including the moon and other celestial bodies, in accordance with international law, including the Charter of the United Nations, in the interest of maintaining international peace and security and promoting international co-operation and understanding.

Article IV

States Parties to the Treaty undertake not to place in orbit around the Earth any objects carrying nuclear weapons or any other kinds of weapons of mass destruction, install such weapons on celestial bodies, or station such weapons in outer space in any other manner.

The Moon and other celestial bodies shall be used by all States Parties to the Treaty exclusively for peaceful purposes. The establishment of military bases, installations and fortifications, the testing of any type of weapons and the conduct of military maneuvers on celestial bodies shall be forbidden. The use of military personnel for scientific research or for any other peaceful purposes shall not be prohibited. The use of any equipment or facility necessary for peaceful exploration of the Moon and other celestial bodies shall also not be prohibited.

Article V

States Parties to the Treaty shall regard astronauts as envoys of mankind in outer space and shall render to them all possible assistance in the event of accident, distress, or emergency landing on the territory of another State Party or on the high seas. When astronauts make such a landing, they shall be safely and promptly returned to the State of registry of their space vehicle.

In carrying on activities in outer space and on celestial bodies, the astronauts of one State Party shall render all possible assistance to the astronauts of other States Parties.

States Parties to the Treaty shall immediately inform the other States Parties to the Treaty or the Secretary-General of the United Nations of any phenomena they discover in outer space, including the Moon and other celestial bodies, which could constitute a danger to the life or health of astronauts.

Article VI

States Parties to the Treaty shall bear international responsibility for national activities in outer space, including the Moon and other celestial bodies, whether such activities are carried on by governmental agencies or by non-governmental entities, and for assuring that national activities are carried out in conformity with the provisions set forth in the present Treaty. The activities of non-governmental entities in outer space, including the Moon and other celestial bodies, shall require authorization and continuing supervision by the appropriate State Party to the Treaty. When activities are carried on in outer space, including the Moon and other celestial bodies, by an international organization, responsibility for compliance with this Treaty shall be borne both by the international organization and by the States Parties to the Treaty participating in such organization.

Article VII

Each State Party to the Treaty that launches or procures the launching of an object into outer space, including the Moon and other celestial bodies, and each State Party from whose territory or facility an object is launched, is internationally liable for damage to another State Party to the Treaty or to its natural or juridical persons by such object or its component parts on the Earth, in air space or in outer space, including the Moon and other celestial bodies.

Article VIII

A State Party to the Treaty on whose registry an object launched into outer space is carried shall retain jurisdiction and control over such object, and over any personnel thereof, while in outer space or on a celestial body. Ownership of objects launched into outer space, including objects landed or constructed on a celestial body, and of their component parts, is not affected by their presence in outer space or on a celestial body or by their return to the Earth. Such objects or component parts found beyond the limits of the State Party to the Treaty on whose registry they are carried shall be returned to that State Party, which shall, upon request, furnish identifying data prior to their return.

Article IX

In the exploration and use of outer space, including the Moon and other celestial bodies, States Parties to the Treaty shall be guided by the principle of co-operation and mutual assistance and shall conduct all their activities in outer space, including the Moon and other celestial bodies, with due regard to the corresponding interests of all other States Parties to the Treaty. States Parties to the Treaty shall pursue studies of outer space, including the Moon and other celestial bodies, and conduct exploration of them so as to avoid their harmful contamination and also adverse changes in the environment of the Earth resulting from the introduction of extraterrestrial matter and, where necessary, shall adopt appropriate measures for this purpose. If a State Party to the Treaty has reason to believe that an activity or experiment planned by it or its nationals in outer space, including the Moon and other celestial bodies, would cause potentially harmful interference with activities of other States Parties in the peaceful exploration and use of outer space, including the Moon and other celestial bodies, it shall undertake appropriate international consultations before proceeding with any such activity or experiment. A State Party to the Treaty which has reason to believe that an activity or experiment planned by another State Party in outer space, including the Moon and other celestial bodies, would cause potentially harmful interference with activities in the peaceful exploration and use of outer space, including the Moon and other celestial bodies, may request consultation concerning the activity or experiment.

Article X

In order to promote international co-operation in the exploration and use of outer space, including the Moon and other celestial bodies, in conformity with the purposes of this Treaty, the States Parties to the Treaty shall consider on a basis of equality any requests by other States Parties to the Treaty to be afforded an opportunity to observe the flight of space objects launched by those States.

The nature of such an opportunity for observation and the conditions under which it could be afforded shall be determined by agreement between the States concerned.

Article XI

In order to promote international co-operation in the peaceful exploration and use of outer space, States Parties to the Treaty conducting activities in outer space, including the Moon and other celestial bodies, agree to inform the Secretary-General of the United Nations as well as the public and the international scientific community, to the greatest extent feasible and practicable, of the nature, conduct, locations and results of such activities. On receiving the said information, the Secretary-General of the United Nations should be prepared to disseminate it immediately and effectively.

Article XII

All stations, installations, equipment and space vehicles on the Moon and other celestial bodies shall be open to representatives of other States Parties to the Treaty on a basis of reciprocity. Such representatives shall give reasonable advance notice of a projected visit, in order that appropriate consultations may be held and that maximum precautions may be taken to assure safety and to avoid interference with normal operations in the facility to be visited.

Article XIII

The provisions of this Treaty shall apply to the activities of States Parties to the Treaty in the exploration and use of outer space, including the Moon and other celestial bodies, whether such activities are carried on by a single State Party to the Treaty or jointly with other States, including cases where they are carried on within the framework of international intergovernmental organizations.

Any practical questions arising in connection with activities carried on by international inter-governmental organizations in the exploration and use of outer space, including the Moon and other celestial bodies, shall be resolved by the States Parties to the Treaty either with the appropriate international organization or with one or more States members of that international organization, which are Parties to this Treaty.

Article XIV

1. This Treaty shall be open to all States for signature. Any State which does not sign this Treaty before its entry into force in accordance with paragraph 3 of this article may accede to it at any time.

2. This Treaty shall be subject to ratification by signatory States. Instruments of ratification and instruments of accession shall be deposited with the Governments of the United States of America, the United Kingdom of Great Britain and Northern Ireland and the Union of Soviet Socialist Republics, which are hereby designated the Depositary Governments.

3. This Treaty shall enter into force upon the deposit of instruments of ratification by five Governments including the Governments designated as Depositary Governments under this Treaty.

4. For States whose instruments of ratification or accession are deposited subsequent to the entry into force of this Treaty, it shall enter into force on the date of the deposit of their instruments of ratification or accession.

5. The Depositary Governments shall promptly inform all signatory and acceding States of the date of each signature, the date of deposit of each instrument of ratification of and accession to this Treaty, the date of its entry into force and other notices.

6. This Treaty shall be registered by the Depositary Governments pursuant to Article 102 of the Charter of the United Nations.

Article XV

Any State Party to the Treaty may propose amendments to this Treaty. Amendments shall enter into force for each State Party to the Treaty accepting the amendments upon their acceptance by a majority of the States Parties to the Treaty and thereafter for each remaining State Party to the Treaty on the date of acceptance by it.

Article XVI

Any State Party to the Treaty may give notice of its withdrawal from the Treaty one year after its entry into force by written notification to the Depositary Governments. Such withdrawal shall take effect one year from the date of receipt of this notification.

Article XVII

This Treaty, of which the English, Russian, French, Spanish and Chinese texts are equally authentic, shall be deposited in the archives of the Depositary Governments. Duly certified copies of this Treaty shall be transmitted by the Depositary Governments to the Governments of the signatory and acceding States.

IN WITNESS WHEREOF the undersigned, duly authorized, have signed this Treaty.

DONE in triplicate, at the cities of Washington, London and Moscow, this twenty-seventh day of January one thousand nine hundred sixty-seven.

Appendix 4

The Registration Convention

**Convention on Registration of Objects Launched into Outer
Space Adopted by the General Assembly of the United
Nations, at New York, on 12 November 1974**

THE STATES PARTIES TO THIS CONVENTION,

RECOGNIZING the common interest of all mankind in furthering the
exploration and use of outer space for peaceful purposes,

RECALLING that the Treaty on principles governing the activities of
States in the exploration and use of outer space, including the moon and
other celestial bodies, of 27 January 1967 affirms that States shall bear
international responsibility for their national activities in outer space and
refers to the State on whose registry an object launched into outer space is
carried,

RECALLING also that the Agreement on the rescue of astronauts, the
return of astronauts and the return of objects launched into outer space of
22 April 1968 provides that a launching authority shall, upon request, furnish identifying data prior to the return of an object it has launched into
outer space found beyond the territorial limits of the launching authority,

RECALLING further that the Convention on international liability for
damage caused by space objects of 29 March 1972 establishes international rules and procedures concerning the liability of launching States for
damage caused by their space objects,

DESIRING, in the light of the Treaty on principles governing the activities
of States in the exploration and use of outer space, including the moon

and other celestial bodies, to make provision for the national registration by launching States of space objects launched into outer space,

DESIRING further that a central register of objects launched into outer space be established and maintained, on a mandatory basis, by the Secretary-General of the United Nations,

DESIRING also to provide for States Parties additional means and procedures to assist in the identification of space objects,

BELIEVING that a mandatory system of registering objects launched into outer space would, in particular, assist in their identification and would contribute to the application and development of international law governing the exploration and use of outer space,

HAVE AGREED ON THE FOLLOWING:

Article I

For the purposes of this Convention:

1. The term "launching State" means:

    1. a State which launches or procures the launching of a space object;

    2. a State from whose territory or facility a space object is launched;

2. The term "space object" includes component parts of a space object as well as its launch vehicle and parts thereof;

3. The term "State of registry" means a launching State on whose registry a space object is carried in accordance with article II.

Article II

1. When a space object is launched into earth orbit or beyond, the launching State shall register the space object by means of an entry in an appropriate registry which it shall maintain. Each launching State shall inform the Secretary-General of the United Nations of the establishment of such a registry.

2. Where there are two or more launching States in respect of any such space object, they shall jointly determine which one of them shall register the object in accordance with paragraph 1 of this article, bearing in mind the provisions of article VIII of the Treaty on principles governing the activities of States in the exploration and use of outer space, including the moon and other celestial bodies, and without prejudice

to appropriate agreements concluded or to be concluded among the launching States on jurisdiction and control over the space object and over any personnel thereof.

3. The contents of each registry and the conditions under which it is maintained shall be determined by the State of registry concerned.

Article III

1. The Secretary-General of the United Nations shall maintain a Register in which the information furnished in accordance with article IV shall be recorded.

2. There shall be full and open access to the information in this Register.

Article IV

1. Each State of registry shall furnish to the Secretary-General of the United Nations, as soon as practicable, the following information concerning each space object carried on its registry:

    1. name of launching State or States;

    2. an appropriate designator of the space object or its registration number;

    3. date and territory or location of launch;

    4. basic orbital parameters, including:

        1. nodal period,

        2. inclination,

        3. apogee,

        4. perigee;

    5. general function of the space object.

2. Each State of registry may, from time to time, provide the Secretary-General of the United Nations with additional information concerning a space object carried on its registry.

3. Each State of registry shall notify the Secretary-General of the United Nations, to the greatest extent feasible and as soon as practicable, of space objects concerning which it has previously transmitted information, and which have been but no longer are in earth orbit.

Article V

Whenever a space object launched into earth orbit or beyond is marked with the designator or registration number referred to in article IV, paragraph 1 (b), or both, the State of registry shall notify the Secretary-General of this fact when submitting the information regarding the space object in accordance with article IV. In such case, the Secretary-General of the United Nations shall record this notification in the Register.

Article VI

Where the application of the provisions of this Convention has not enabled a State Party to identify a space object which has caused damage to it or to any of its natural or juridical persons, or which may be of a hazardous or deleterious nature, other States Parties, including in particular States possessing space monitoring and tracking facilities, shall respond to the greatest extent feasible to a request by that State Party, or transmitted through the Secretary-General on its behalf, for assistance under equitable and reasonable conditions in the identification of the object. A State Party making such a request shall, to the greatest extent feasible, submit information as to the time, nature and circumstances of the events giving rise to the request. Arrangements under which such assistance shall be rendered shall be the subject of agreement between the parties concerned.

Article VII

1. In this Convention, with the exception of articles VIII to XII inclusive, references to States shall be deemed to apply to any international inter-governmental organization which conducts space activities if the organization declares its acceptance of the rights and obligations provided for in this Convention and if a majority of the States members of the organization are States Parties to this Convention and to the Treaty on principles governing the activities of States in the exploration and use of outer space, including the moon and other celestial bodies.

2. States members of any such organization which are States Parties to this Convention shall take all appropriate steps to ensure that the organization makes a declaration in accordance with paragraph 1 of this article.

Article VIII

1. This Convention shall be open for signature by all States at United Nations Headquarters in New York. Any State which does not sign this

Convention before its entry into force in accordance with paragraph 3 of this article may accede to it at any time.

2. This Convention shall be subject to ratification by signatory States. Instruments of ratification and instruments of accession shall be deposited with the Secretary-General of the United Nations.

3. This Convention shall enter into force among the States which have deposited instruments of ratification on the deposit of the fifth such instrument with the Secretary-General of the United Nations.

4. For States whose instruments of ratification or accession are deposited subsequent to the entry into force of this Convention, it shall enter into force on the date of the deposit of their instruments of ratification or accession.

5. The Secretary-General shall promptly inform all signatory and acceding States of the date of each signature, the date of deposit of each instrument of ratification of and accession to this Convention, the date of its entry into force and other notices.

Article IX

Any State Party to this Convention may propose amendments to the Convention. Amendments shall enter into force for each State Party to the Convention accepting the amendments upon their acceptance by a majority of the States Parties to the Convention and thereafter for each remaining State Party to the Convention on the date of acceptance by it.

Article X

Ten years after the entry into force of this Convention, the question of the review of the Convention shall be included in the provisional agenda of the United Nations General Assembly in order to consider, in the light of past application of the Convention, whether it requires revision. However, at any time after the Convention has been in force for five years, at the request of one third of the States Parties to the Convention and with the concurrence of the majority of the States Parties, a conference of the States Parties shall be convened to review this Convention. Such review shall take into account in particular any relevant technological developments, including those relating to the identification of space objects.

Article XI

Any State Party to this Convention may give notice of its withdrawal from the Convention one year after its entry into force by written notification to the Secretary-General of the United Nations. Such withdrawal shall take effect one year from the date of receipt of this notification.

Article XII

The original of this Convention, of which the Arabic, Chinese, English, French, Russian and Spanish texts are equally authentic, shall be deposited with the Secretary-General of the United Nations, who shall send certified copies thereof to all signatory and acceding States.

IN WITNESS WHEREOF the undersigned, being duly authorized thereto by their respective Governments, have signed this Convention, opened for signature at New York on the fourteenth day of January one thousand nine hundred and seventy-five.

Appendix 5

General Assembly Resolution 34/68

**Resolution Adopted by The General Assembly 34/68.
Agreement Governing the Activities of States on the
Moon and Other Celestial Bodies**

The General Assembly,

Reaffirming the importance of international cooperation in the field of the exploration and peaceful uses of outer space, including the moon and other celestial bodies, and of promoting the rule of law in this field of human endeavour,

Recalling its resolution 2779 (XXVI) of 29 November 1971, in which it requested the Committee on the Peaceful Uses of Outer Space and its Legal Subcommittee to consider the question of the elaboration of a draft international treaty concerning the moon, as well as its resolution 2915 (XXVII) of 9 November 1972, 3182 (XXVIII) of 18 December 1973, 3234 (XXIX) of 12 November 1974, 3388 (XXX) of 18 November 1975, 31/8 of 8 November 1976, 32/196 A of 20 December 1977 and 33/16 of 10 November 1978, in which it, inter alia, encouraged the elaboration of the draft treaty relating to the moon.

Recalling, in particular, that in resolution 33/16 it endorsed the recommendation of the Committee on the Peaceful Uses of Outer Space that the Legal Subcommittee at its eighteenth session should continue as a matter of priority its efforts to complete the draft treaty relating to the moon,

Having considered the relevant part of the report of the Committee on the Peaceful Uses of Outer Space, in particular paragraphs 62, 63 and 65.

Noting with satisfaction that the Committee on the Peaceful Uses of Outer Space, on the basis of the deliberations and recommendations of the Legal Subcommittee, has completed the text of the draft Agreement Governing the Activities of States on the Moon and Other Celestial Bodies,

Having considered the text of the draft Agreement Governing the Activities of States on the Moon and Other Celestial Bodies,[1]

1. *Commends* the Agreement Governing the Activities of States on the Moon and Other Celestial Bodies, the text of which is annexed to the present resolution;

2. *Requests* the Secretary-General to open the Agreement for signature and ratification at the earliest possible date;

3. *Expresses* its hope for the widest possible adherence to this Agreement.

89th plenary meeting,
5 December 1979.

Appendix 6

The Moon Agreement

Agreement Governing the Activities of States on the Moon and Other Celestial Bodies

THE STATES PARTIES TO THIS CONVENTION,

NOTING the achievements of States in the exploration and use of the moon and other celestial bodies,

RECOGNIZING that the moon, as a natural satellite of the earth, has an important role to play in the exploration of outer space,

DETERMINED to promote on the basis of equality the further development of co-operation among States in the exploration and use of the moon and other celestial bodies,

DESIRING to prevent the moon from becoming an area of international conflict,

BEARING in mind the benefits which may be derived from the exploitation of the natural resources of the moon and other celestial bodies,

RECALLING the Treaty on Principles Governing the Activities of States in the Exploration and Use of Outer Space, including the Moon and Other Celestial Bodies, the Agreement on the Rescue of Astronauts, the Return of Astronauts and the Return of Objects Launched into Outer Space, the Convention on International Liability for Damage Caused by Space Objects, and the Convention on Registration of Objects Launched into Outer Space,

TAKING INTO ACCOUNT the need to define and develop the provisions of these international instruments in relation to the moon and other celestial bodies, having regard to further progress in the exploration and use of outer space,

HAVE AGREED ON THE FOLLOWING:

Article 1

1. The provisions of this Agreement relating to the moon shall also apply to other celestial bodies within the solar system, other than the earth, except in so far as specific legal norms enter into force with respect to any of these celestial bodies.

2. For the purposes of this Agreement reference to the moon shall include orbits around or other trajectories to or around it.

3. This Agreement does not apply to extraterrestrial materials which reach the surface of the earth by natural means.

Article 2

All activities on the moon, including its exploration and use, shall be carried out in accordance with international law, in particular the Charter of the United Nations, and taking into account the Declaration on Principles of International Law concerning Friendly Relations and Co-operation among States in accordance with the Charter of the United Nations, adopted by the General Assembly on 24 October 1970, in the interest of maintaining international peace and security and promoting international co-operation and mutual understanding, and with due regard to the corresponding interests of all other States Parties.

Article 3

1. The moon shall be used by all States Parties exclusively for peaceful purposes.

2. Any threat or use of force or any other hostile act or threat of hostile act on the moon is prohibited. It is likewise prohibited to use the moon in order to commit any such act or to engage in any such threat in relation to the earth, the moon, spacecraft, the personnel of spacecraft or man-made space objects.

3. States Parties shall not place in orbit around or other trajectory to or around the moon objects carrying nuclear weapons or any other kinds

of weapons of mass destruction or place or use such weapons on or in the moon.

4. The establishment of military bases, installations and fortifications, the testing of any type of weapons and the conduct of military manoeuvres on the moon shall be forbidden. The use of military personnel for scientific research or for any other peaceful purposes shall not be prohibited. The use of any equipment or facility necessary for peaceful exploration and use of the moon shall also not be prohibited.

Article 4

1. The exploration and use of the moon shall be the province of all mankind and shall be carried out for the benefit and in the interests of all countries, irrespective of their degree of economic or scientific development. Due regard shall be paid to interests of present and future generations as well as to the need to promote higher standards of living conditions of economic and social progress and development in accordance with the Charter of the United Nations.

2. States Parties shall be guided by the principle of co-operation and mutual assistance in all their activities concerning the exploration and use of the moon. International co-operation in pursuance of this Agreement should be as wide as possible and may take place on a multilateral basis, on a bilateral basis or through international inter-governmental organizations.

Article 5

1. States Parties shall inform the Secretary-General of the United Nations as well as the public and the international scientific community, to the greatest extent feasible and practicable, of their activities concerned with the exploration and use of the moon. Information on the time, purposes, locations, orbital parameters and duration shall be given in respect of each mission to the moon as soon as possible after launching, while information on the results of each mission, including scientific results, shall be furnished upon completion of the mission. In the case of a mission lasting more than sixty days, information on conduct of the mission including any scientific results, shall be given periodically, at thirty-day intervals. For missions lasting more than six months, only significant additions to such information need be reported thereafter.

2. If a State Party becomes aware that another State Party plans to operate simultaneously in the same area of or in the same orbit around or

trajectory to or around the moon, it shall promptly inform the other State of the timing of and plans for its own operations.

3. In carrying out activities under this Agreement, States Parties shall promptly inform the Secretary-General, as well as the public and the international scientific community, of any phenomena they discover in outer space, including the moon, which could endanger human life or health, as well as of any indication of organic life.

Article 6

1. There shall be freedom of scientific investigation on the moon by all States Parties without discrimination of any kind, on the basis of equality and in accordance with international law.

2. In carrying out scientific investigations and in furtherance of the provisions of this Agreement, the States Parties shall have the right to collect on and remove from the moon samples of its mineral and other substances. Such samples shall remain at the disposal of those States Parties which caused them to be collected and may be used by them for scientific purposes. States Parties shall have regard to the desirability of making a portion of such samples available to other interested States Parties and the international scientific community for scientific investigation. States Parties may in the course of scientific investigations also use mineral and other substances of the moon in quantities appropriate for the support of their missions.

3. States Parties agree on the desirability of exchanging scientific and other personnel on expeditions to or installations on the moon to the greatest extent feasible and practicable.

Article 7

1. In exploring and using the moon, States Parties shall take measures to prevent the disruption of the existing balance of its environment, whether by introducing adverse changes in that environment, by its harmful contamination through the introduction of extra-environmental matter or otherwise. States Parties shall also take measures to avoid harmfully affecting the environment of the earth through the introduction of extraterrestrial matter or otherwise.

2. States Parties shall inform the Secretary-General of the United Nations of the measures being adopted by them in accordance with paragraph 1 of this article and shall also, to the maximum extent feasible, notify

him in advance of all placements by them of radio-active materials on the moon and of the purposes of such placements.

3. States Parties shall report to other States Parties and to the Secretary-General concerning areas of the moon having special scientific interest in order that, without prejudice to the rights of other States Parties, consideration may be given to the designation of such areas as international scientific preserves for which special protective arrangements are to be agreed upon in consultation with the competent bodies of the United Nations.

Article 8

1. States Parties may pursue their activities in the exploration and use of the moon anywhere on or below its surface, subject to the provisions of this Agreement.

2. For these purposes States Parties may, in particular:

   1. Land their space objects on the moon and launch them from the moon;

   2. Place their personnel, space vehicles, equipment, facilities, stations and installations anywhere on or below the surface of the moon.

   Personnel, space vehicles, equipment, facilities, stations and installations may move or be moved freely over or below the surface of the moon.

3. Activities of States Parties in accordance with paragraphs 1 and 2 of this article shall not interfere with the activities of other States Parties on the moon. Where such interference may occur, the States Parties concerned shall undertake consultations in accordance with article 15, paragraphs 2 and 3, of this Agreement.

Article 9

1. States Parties may establish manned and unmanned stations on the moon. A State Party establishing a station shall use only that area which is required for the needs of the station and shall immediately inform the Secretary-General of the United Nations of the location and purposes of that station. Subsequently, at annual intervals that State shall likewise inform the Secretary-General whether the station continues in use and whether its purposes have changed.

2. Stations shall be installed in such a manner that they do not impede the free access to all areas of the moon of personnel, vehicles and equipment

of other States Parties conducting activities on the moon in accordance with the provisions of this Agreement or of article I of the Treaty of Principles Governing the Activities of States in the Exploration and Use of Outer Space, including the Moon and other Celestial Bodies.

Article 10

1. States Parties shall adopt all practicable measures to safeguard the life and health of persons on the moon. For this purpose they shall regard any person on the moon as an astronaut within the meaning of article V of the Treaty on Principles Governing the Activities of States on the Exploration and Use of Outer Space, including the Moon and Other Celestial Bodies and as part of the personnel of a spacecraft within the meaning of the Agreement on the Rescue of Astronauts, the Return of Astronauts and the Return of Objects Launched into Outer Space.

2. States Parties shall offer shelter in their stations, installations, vehicles and other facilities to persons in distress on the moon.

Article 11

1. The moon and its natural resources are the common heritage of mankind, which finds its expression in the provisions of this Agreement, in particular in paragraph 5 of this article.

2. The moon is not subject to national appropriation by any claim of sovereignty, by means of use or occupation, or by any other means.

3. Neither the surface nor the subsurface of the moon, nor any part thereof or natural resources in place, shall become property of any State, international intergovernmental or non-governmental organization, national organization or non-governmental entity or of any natural person. The placement of personnel, space vehicles, equipment, facilities, stations and installations on or below the surface of the moon, including structures connected with its surface or subsurface, shall not create a right of ownership over the surface or the subsurface of the moon or any areas thereof. The foregoing provisions are without prejudice to the international regime referred to in paragraph 5 of this article.

4. States Parties have the right to exploration and use of the moon without discrimination of any kind, on the basis of equality and in accordance with international law and the provisions of this Agreement.

5. States Parties to this Agreement hereby undertake to establish an international regime, including appropriate procedures, to govern the exploitation of the natural resources of the moon as such exploitation is about to become feasible. This provision shall be implemented in accordance with article 18 of this Agreement.

6. In order to facilitate the establishment of the international regime referred to in paragraph 5 of this article, States Parties shall inform the Secretary-General of the United Nations as well as the public and the international scientific community, to the greatest extent feasible and practicable, of any natural resources they may discover on the moon.

7. The main purposes of the international regime to be established shall include:

   1. The orderly and safe development of the natural resources of the moon;

   2. The rational management of those resources;

   3. The expansion of opportunities in the use of those resources;

   4. An equitable sharing by all States Parties in the benefits derived from those resources, whereby the interests and needs of the developing countries, as well as the efforts of those countries which have contributed either directly or indirectly to the exploration of the moon, shall be given special consideration.

8. All the activities with respect to the natural resources of the moon shall be carried out in a manner compatible with the purposes specified in paragraph 7 of this article and the provisions of article 6, paragraph 2, of this Agreement.

Article 12

1. States Parties shall retain jurisdiction and control over their personnel, space vehicles, equipment, facilities, stations and installations on the moon. The ownership of space vehicles, equipment, facilities, stations and installations shall not be affected by their presence on the moon.

2. Vehicles, installations and equipment or their component parts found in places other than their intended location shall be dealt with in accordance with article 5 of the Agreement on the Rescue of Astronauts, the Return of Astronauts and the Return of Objects Launched into Outer Space.

3. In the event of an emergency involving a threat to human life, States Parties may use the equipment, vehicles, installations, facilities or supplies of other States Parties on the moon. Prompt notification of such use shall be made to the Secretary-General of the United Nations or the State Party concerned.

Article 13

A State Party which learns of the crash landing, forced landing or other unintended landing on the moon of a space object, or its component parts, that were not launched by it, shall promptly inform the launching State Party and the Secretary-General of the United Nations.

Article 14

1. States Parties to this Agreement shall bear international responsibility for national activities on the moon, whether such activities are carried out by governmental agencies or by non-governmental entities, and for assuring that national activities are carried out in conformity with the provisions of this Agreement. States Parties shall ensure that non-governmental entities under their jurisdiction shall engage in activities on the moon only under the authority and continuing supervision of the appropriate State Party.

2. States Parties recognize that detailed arrangements concerning liability for damage caused on the moon, in addition to the provisions of the Treaty on Principles Governing the Activities of States in the Exploration and Use of Outer Space, including the Moon and Other Celestial Bodies and the Convention on International Liability for Damage Caused by Space Objects, may become necessary as a result of more extensive activities on the moon. Any such arrangements shall be elaborated in accordance with the procedure provided for in article 18 of this Agreement.

Article 15

1. Each State Party may assure itself that the activities of other States Parties in the exploration and use of the moon are compatible with the provisions of this Agreement. To this end, all space vehicles, equipment, facilities, stations and installations on the moon shall be open to other States Parties. Such States Parties shall give reasonable advance notice of a projected visit, in order that appropriate consultations may be held and that maximum precautions may be taken to assure safety

and to avoid interference with normal operations in the facility to be visited. In pursuance of this article, any State Party may act on its own behalf or with the full or partial assistance of any other State Party or through appropriate international procedures within the framework of the United Nations and in accordance with the Charter.

2. A State Party which has reason to believe that another State Party is not fulfilling the obligations incumbent upon it pursuant to this Agreement or that another State Party is interfering with the rights which the former State Party has under this Agreement may request consultations with that State Party. A State Party receiving such a request shall enter into such consultations without delay. Any other State Party which requests to do so shall be entitled to take part in the consultations. Each State Party participating in such consultations shall seek a mutually acceptable resolution of any controversy and shall bear in mind the rights and interests of all States Parties. The Secretary-General of the United Nations shall be informed of the results of the consultations and shall transmit the information received to all States Parties concerned.

3. If the consultations do not lead to a mutually acceptable settlement which has due regard for the rights and interests of all the States Parties, the parties concerned shall take all measures to settle the dispute by other peaceful means of their choice and appropriate to the circumstances and the nature of the dispute. If difficulties arise in connexion with the opening of consultations or if consultations do not lead to a mutually acceptable settlement, any State Party may seek the assistance of the Secretary-General, without seeking the consent of any other State Party concerned, in order to resolve the controversy. A State Party which does not maintain diplomatic relations with another State Party concerned shall participate in such consultations, at its choice, either itself or through another State Party or the Secretary-General as intermediary.

Article 16

With the exception of articles 17 to 21, references in this Agreement to States shall be deemed to apply to any international intergovernmental organization which conducts space activities if the organization declares its acceptance of the rights and obligations provided for in this Agreement and if a majority of the States members of the organization are States Parties to this Agreement and to the Treaty on Principles Governing the Activities of States in the Exploration and Use of Outer Space, including the Moon and Other Celestial Bodies. States members of any such organization which are States Parties to this Agreement shall take all appropriate steps to ensure that the organization makes a declaration in accordance with the provisions of this article.

Article 17

Any State Party to this Agreement may propose amendments to the Agreement. Amendments shall enter into force for each State Party to the Agreement accepting the amendments upon their acceptance by a majority of the States Parties to the Agreement and thereafter for each remaining State Party to the Agreement on the date of acceptance by it.

Article 18

Ten years after the entry into force of this Agreement, the question of the review of the Agreement shall be included in the provisional agenda of the General Assembly of the United Nations in order to consider, in the light of past application of the Agreement, whether it requires revision. However, at any time after the Agreement has been in force for five years, the Secretary-General of the United Nations, as depository, shall, at the request of one third of the States Parties to the Agreement and with the concurrence of the majority of the States Parties, convene a conference of the States Parties to review this Agreement. A review conference shall also consider the question of the implementation of the provisions of article 11, paragraph 5, on the basis of the principle referred to in paragraph 1 of that article and taking into account in particular any relevant technological developments.

Article 19

1. This Agreement shall be open for signature by all States at United Nations Headquarters in New York.

2. This Agreement shall be subject to ratification by signatory States. Any State which does not sign this Agreement before its entry into force in accordance with paragraph 3 of this article may accede to it at any time. Instruments of ratification or accession shall be deposited with the Secretary-General of the United Nations.

3. This Agreement shall enter into force on the thirtieth day following the date of deposit of the fifth instrument of ratification.

4. For each State depositing its instrument of ratification or accession after the entry into force of this Agreement, it shall enter into force on the thirtieth day following the date of deposit of any such instrument.

5. The Secretary-General shall promptly inform all signatory and acceding States of the date of each signature, the date of deposit of each instrument of ratification or accession to this Agreement, the date of its entry into force and other notices.

Article 20

Any State Party to this Agreement may give notice of its withdrawal from the Agreement one year after its entry into force by written notification to the Secretary-General of the United Nations. Such withdrawal shall take effect one year from the date of receipt of this notification.

Article 21

The original of this Agreement, of which the Arabic, Chinese, English, French, Russian and Spanish texts are equally authentic, shall be deposited with the Secretary-General of the United Nations, who shall send certified copies thereof to all signatory and acceding States.

IN WITNESS WHEREOF the undersigned, being duly authorized thereto by their respective Governments, have signed this Agreement, opened for signature at New York on 18 December 1979.

Appendix 7

The COPUOS Understandings of the Moon Agreement

62. Several suggestions were made to amend article I, paragraph 1. However, after an extensive discussion of the matter, it was agreed not to amend the Austrian text but to include in the report of the Committee a statement reflecting the Committee's understanding of the interpretation that should be given to article I, paragraph 1. That understanding is as follows:

   "The committee agreed that by virtue of article I, paragraph 1, the principle contained in article XI, paragraph 1 would also apply to celestial bodies in the solar system other than the Earth and to its natural resources."

63. Following a suggestion for clarification of article I, paragraph 2, the committee agreed that the trajectories and orbits mentioned in article I, paragraph 2, do not include trajectories and orbits of space objects in Earth orbits only and trajectories of space objects between the Earth and such orbits.

65. Following a suggestion for further clarification of article VII, the committee agreed that article VII is not intended to result in prohibiting the exploitation of natural resources which may be found on celestial bodies other than the Earth but, rather, that such exploitation will be carried out in such manner as to minimize any disruption or adverse effects to the existing balance of the environment.

Appendix 8

The ABA Section of Intl. Law Resolution on the Moon Agreement

Be it resolved that the American Bar Association favors the signature and ratification by the United States of the Agreement Governing the Activities of States on the Moon and Other Celestial Bodies, and urges the Senate to give its advice and consent to ratification, subject to the inclusion of the following understandings and declarations in the instrument of ratification:

(a) It is the understanding of the United States that no provision in this Agreement constrains the existing right of governmental or authorized non-governmental entities to explore and use the resources of the moon or other celestial body, including the right to develop and exploit these resources for commercial or other purposes. In addition, it is the understanding of the United States that nothing in this Agreement in any way diminishes or alters the right of the United States to determine how it shares the benefits derived from exploitation by or under the authority of the United States of natural resources of the moon or other celestial bodies;

(b) Natural resources extracted, removed or actually utilized by or under the authority of a State Party to this Agreement are subject to the exclusive control of, and may be considered as the property of, the State Party or other entity responsible for their extraction, removal or utilization;

(c) The meaning of the term "common heritage of mankind" is to be based on the provisions of this Agreement, and not on the use or interpretation of that term in any other context. Recognition by the United States that the moon and its natural resources are the common heritage of mankind constitutes recognition (A) that all States have equal rights to explore and use the moon and its natural resources, and (B) that no State or other entity has an exclusive right of ownership, property or appropriation over the moon, over any area of the surface or subsurface of the moon, or over its natural resources in place. In this context, the United States notes that, in accordance with Articles XII and XV of this Agreement, States Parties retain exclusive jurisdiction and control over their facilities, stations and installations on the moon, and that other States Parties are obligated to avoid interference with normal operations on such facilities.

(d) Acceptance by the United States of an obligation to undertake in the future good faith negotiation with other States Parties of an international regime to govern exploitation of the natural resources of the moon in no way prejudices the existing right of the United States to exploit or authorize the exploitation of those natural resources. No moratorium on such exploitation is intended or required by this Agreement. The United States recognizes that States Parties to this Agreement are obligated to act in a manner compatible with the provisions of Article VI(2) and the purposes specified in Article XI(7); however, the United States reserves to itself the right and authority to determine the standards for such compatibility unless and until the United States becomes a party to a future resources exploitation regime. In addition, acceptance of the obligation to join in good faith negotiation of such a regime in no way constitutes acceptance of any particular provisions which may be included in such a regime; nor does it constitute an obligation to become a Party to such a regime regardless of its contents.

Appendix 9

The Declaration on International Cooperation

Declaration on International Cooperation in the Exploration and Use of Outer Space for the Benefit and in the Interest of all States, Taking into Particular Account the Needs of Developing Countries

The General Assembly,

Having considered the report of the Committee on the Peaceful Uses of Outer Space on the work of its thirty-ninth session and the text of the Declaration on International Cooperation in the Exploration and Use of Outer Space for the Benefit and in the Interest of All States, Taking into Particular Account the Needs of Developing Countries, as approved by the Committee and annexed to its report,

Bearing in mind the relevant provisions of the Charter of the United Nations,

Recalling notably the provisions of the Treaty on the Principles Governing the Activities of States in the Exploration and Use of Outer Space, including the Moon and Other Celestial Bodies,

Recalling also its relevant resolutions relating to activities in outer space,

Bearing in mind the recommendations of the Second United Nations Conference on the Exploration and Peaceful Uses of Outer Space, and of other international conferences relevant in this field,

Recognizing the growing scope and significance of international cooperation among States and between States and international organizations in the exploration and use of outer space for peaceful purposes,

Considering experiences gained in international cooperative ventures,

Convinced of the necessity and the significance of further strengthening international cooperation in order to reach a broad and efficient collaboration in this field for the mutual benefit and in the interest of all parties involved,

Desirous of facilitating the application of the principle that the exploration and use of outer space, including the Moon and other celestial bodies, shall be carried out for the benefit and in the interest of all countries, irrespective of their degree of economic or scientific development, and shall be the province of all mankind,

Adopts the Declaration on International Cooperation in the Exploration and Use of Outer Space for the Benefit and in the Interest of All States, Taking into Particular Account the Needs of Developing Countries, set forth in the annex to the present resolution.

Annex

Declaration on International Cooperation in the Exploration and Use of Outer Space for the Benefit and in the Interest of all States, Taking into Particular Account the Needs of Developing Countries

1. International cooperation in the exploration and use of outer space for peaceful purposes (hereafter "international cooperation") shall be conducted in accordance with the provisions of international law, including the Charter of the United Nations and the Treaty on the Principles Governing the Activities of States in the Exploration and Use of Outer Space, including the Moon and Other Celestial Bodies. It shall be carried out for the benefit and in the interest of all States, irrespective of their degree of economic, social or scientific and technological development, and shall be the province of all mankind. Particular account should be taken of the needs of developing countries.

2. States are free to determine all aspects of their participation in international cooperation in the exploration and use of outer space on an equitable and mutually acceptable basis. Contractual terms in such cooperative ventures should be fair and reasonable and they should be in full compliance with the legitimate rights and interests of the parties concerned as, for example, with intellectual property rights.

3. All States, particularly those with relevant space capabilities and with programmes for the exploration and use of outer space, should

contribute to promoting and fostering international cooperation on an equitable and mutually acceptable basis. In this context, particular attention should be given to the benefit for and the interests of developing countries and countries with incipient space programmes stemming from such international cooperation conducted with countries with more advanced space capabilities.

4. International cooperation should be conducted in the modes that are considered most effective and appropriate by the countries concerned, including, inter alia, governmental and non-governmental; commercial and non-commercial; global, multilateral, regional or bilateral; and international cooperation among countries in all levels of development.

5. International cooperation, while taking into particular account the needs of developing countries, should aim, inter alia, at the following goals, considering their need for technical assistance and rational and efficient allocation of financial and technical resources:

    (a) Promoting the development of space science and technology and of its applications;

    (b) Fostering the development of relevant and appropriate space capabilities in interested States;

    (c) Facilitating the exchange of expertise and technology among States on a mutually acceptable basis.

6. National and international agencies, research institutions, organizations for development aid, and developed and developing countries alike should consider the appropriate use of space applications and the potential of international cooperation for reaching their development goals.

7. The Committee on the Peaceful Uses of Outer Space should be strengthened in its role, among others, as a forum for the exchange of information on national and international activities in the field of international cooperation in the exploration and use of outer space.

8. All States should be encouraged to contribute to the United Nations Programme on Space Applications and to other initiatives in the field of international cooperation in accordance with their space capabilities and their participation in the exploration and use of outer space.

Index

About the Author

Thomas Gangale is Executive Director of OPS-Alaska, an aerospace think tank in Petaluma, California. He served as an Air Force space program engineer for satellite programs involving strategic arms control verification, the Strategic Defense Initiative, and Space Shuttle payloads. He holds degrees in aerospace engineering from the University of Southern California and in international relations from San Francisco State University. His work on property rights and the international law of outer space has been briefed to senior NASA leaders. He has published numerous articles on space law, astronautics, and planetary timekeeping systems in such journals as *Planetary and Space Science, Journal of Aerospace, American Institute of Aeronautics and Astronautics, Annals of Air and Space Law,* and *Journal of the British Interplanetary Society.*